Heterogeneous Photochemical Electron Transfer

Author

Michael Grätzel, Ph.D.

Professor of Chemistry
Ecole Polytechnique Fédérale de Lausanne
Lausanne, Switzerland

CRC Press
Taylor & Francis Group
Boca Raton London New York

CRC Press is an imprint of the
Taylor & Francis Group, an **informa** business

First published 1989 by CRC Press
Taylor & Francis Group
6000 Broken Sound Parkway NW, Suite 300
Boca Raton, FL 33487-2742

Reissued 2018 by CRC Press

© 1989 by CRC Press, Inc.
CRC Press is an imprint of Taylor & Francis Group, an Informa business

No claim to original U.S. Government works

Library of Congress Cataloging-in-Publication Data

Grätzel, Michael.
    Heterogeneous photochemical electron transfer.

Bibliography: p.
    Includes index.
    I. Photochemistry.   2. Oxidation-reduction reaction.
I. Title.
QD716.095G73   1988   541.3'5   87-34117
ISBN 0-8493-5968-6

A Library of Congress record exists under LC control number: 87034117

ISBN 13: 978-1-315-89410-2 (hbk)
ISBN 13: 978-1-351-07320-2 (ebk)

Visit the Taylor & Francis Web site at http://www.taylorandfrancis.com and the
CRC Press Web site at http://www.crcpress.com

DEDICATION

To my parents, my wife, Carole, and my children, Chauncey, Aimie-Lynn, and Liliane, with love.

ACKNOWLEDGMENTS

I wish to express my deep gratitude to my co-workers for their important contribution to many of the research results discussed in this book. In particular, I am indebted to Drs. G. Rothenberger, P. P. Infelta, and J. Moser for numerous helpful discussions, to Francine Duriaux for her skillful help with the art work, and to Nelly Gourdou, Anita Magnenat, and Kristine Verhamme for their valuable assistance with the typing of the manuscript. Last, but not least, I wish to thank Professors Maurice Cosandey and Bernard Vittoz, both Presidents of the Ecole Polytechnique Fédérale de Lausanne during my tenure, for their generous support and constant encouragement.

Michael Grätzel, Ph.D.
November 1987

PREFACE

Over the past few years, the field of thermal and photoinduced electron-transfer reactions in heterogeneous systems has witnessed an astonishing number of exciting new discoveries. These concern such important problems as the unraveling of the intricate mechanism of the primary charge separation process in photosynthesis and other life-sustaining biological processes, e.g., mitochondrial respiration. The development of artificial photosynthetic systems to harvest sunlight and convert it into electricity or chemical fuels is rapidly advancing. For example, at present, the molecular engineering of organized molecular systems to effect light-induced charge separation constitutes a rapidly growing research front. Among the different structural units that are in the focus of present investigations, colloidal semiconductors exhibit particularly intriguing properties. Investigations of redox processes in these systems have yielded a wealth of information on the dynamics of charge transfer reactions within the semiconductors and across the interface separating it from the solution. For the interpretation of these processes, novel kinetic models were conceived taking into account the unique features of microheterogeneous reactions systems. In parallel, the elementary theory of electron-transfer reactions in condensed systems has progressed significantly over the past few years.

The present monograph attempts to unify these diverse and exciting new developments within a common framework. First, the physical principles underlying heterogeneous electron-transfer processes are outlined in a concise way and are compared to the homogeneous counterpart. This analysis includes the notion of the Fermi level in liquids and solids as well as the distribution of electronic energy levels in solids and liquids. A comparison is made between the salient kinetic features of homogeneous and heterogeneous electron-transfer reactions. This establishes the basis for the subsequent treatment of the transduction of excitation energy and photo-initiated electron transfer in organized molecular assemblies, such as micelles, vesicles, and monolayers. Transmembrane redox processes are critically reviewed. Particular attention is given to semiconductor electrodes and particles. This includes a discussion of quantum size effects, the nature of space charge layers as well as surface states and the dynamics of charge carrier-induced redox reactions at the semiconductor solution interface. These processes are of fundamental importance in such diverse fields as photochromism, electrochromic displays, electroreprography and photography, information storage, photocatalysis, photodegradation of paints, and solar energy conversion.

Our book should be helpful to researchers working in these and related domains as well as to postgraduate students. We would be delighted if it would contribute to the progress of the study of heterogeneous photochemical electron-transfer reactions.

THE AUTHOR

Michael Grätzel, Ph.D., is a Professor at the Institute of Physical Chemistry at the Swiss Federal Institute of Technology in Lausanne, Switzerland. He was Director of the Institute from 1980 to 1982 and served as Head of the Chemistry Department from 1983 to 1985.

Dr. Grätzel received his higher education at two universities in West Germany. He carried out his undergraduate studies in Chemistry at the Free University of Berlin and was awarded a fellowship of the Studienstiftung des Deutschen Volkes to complete his Diploma and Doctor's Degree. During his diploma thesis, he worked on the crystallization kinetics of polymers which he investigated at the Fritz-Haber-Institute of the Max Planck Society.

In 1968 he entered the Technical University of Berlin for graduate studies. He performed his research at the Hahn-Meitner-Institute where he investigated the short-lived intermediates formed by one-electron oxidation and reduction of nitrate, nitrite, and nitric oxide in aqueous solution. Important thermodynamic and kinetic properties of several key intermediates, such as the dimerization constant of NO_2 in water and the hydrolysis rate of N_2O_4, were determined for the first time during these studies. His thesis work was published as five papers in a scientific journal, *Berichte der Bunsengesellschaft fuer Physikalische Chemie*.

Dr. Grätzel joined the Chemistry Department and Radiation Laboratory of the University of Notre Dame in Notre Dame, Ind. as a Petroleum Research Foundation Fellow in 1972. His research there was concerned with the application of photochemical and NMR techniques, to explore the structure of micelles and membranes. In addition, laser photoionization studies in surfactant solutions were performed and electron-transfer reactions across the lipid water interface investigated.

During 1975 and 1976 he returned to Berlin as a staff scientist at the Hahn-Meitner-Institute where he began his work on photo-induced redox reactions in organized molecular assemblies. He submitted his Habilitationsschrift to the Free University of Berlin and was awarded the venia legendi for Physical Chemistry.

In 1977 he was appointed Associate Professor of the Department of Chemistry at the Swiss Federal Institute of Technology (Ecole Polytechnique Fédérale de Lausanne) in Lausanne, Switzerland. In 1981 he was promoted to the rank of Full Professor. In Lausanne, Dr. Grätzel established a new research program in heterogeneous photochemical electron-transfer reactions, redox catalysis, and photocatalysis. His laboratory has been active in developing functional micellar assemblies for light-induced charge separation and redox catalysts for the oxidation of water to oxygen and reduction of water to hydrogen. It pioneered the investigation of charge transfer processes in colloidal semiconductor dispersions. Furthermore, the laboratory initiated the development of novel kinetic concepts to interpret the dynamics of fast reactions in microheterogeneous systems. These systems have been applied in the photochemical cleavage of water and hydrogen sulfide, the destruction of environmental pollutants, the photocatalytic conversion of CO_2 to methane, and the activation of methane. More recently, the laboratory research has been extended to high surface area semiconductor oxide films with fractal dimensions. Combining these layers with suitable sensitizers has led to the discovery of highly efficient light harvesting devices which are employed to generate electricity from sunlight.

Dr. Grätzel regularly performs research as a guest scientist at the Solar Energy Research Institute in Golden, Colo. and has been invited as a Visiting Professor by the University of California at Berkeley and the Lawrence Berkeley Laboratory. He has held honorary lectureships from the British Council and the University of Texas at Austin and was awarded a fellowship of the Japanese Society for the Promotion of Science.

Dr. Grätzel is the author of over 260 publications as well as the editor of the book *Energy Resources Through Photosynthesis and Catalysis* which was published by Academic Press in 1983. He is a member of the Editorial Board of the *Journal of Molecular Catalysis* and a member of the Advisory Board of the recent CRC series on *Advances in Photochemistry and Photophysics*. At present, he is President of the Chemical Section of the Société Vaudoise de Sciences Naturelles.

TABLE OF CONTENTS

Chapter 1

THEORETICAL ASPECTS OF ELECTRON-TRANSFER REACTIONS

I. GENERAL FEATURES OF ELECTRON-TRANSFER REACTIONS

The transfer of an electron from one site to another in a molecule, as well as between molecules or redox centers located in two different phases, is one of the most fundamental and ubiquitous processes in chemistry and biology. For decades researchers have investigated this reaction from both the theoretical and experimental points of view.[1] There are two crucial questions which need to be dealt with. First, it is important to establish how far an electron can travel between redox sites in a single, concerted reaction. In other words, what role does the distance separating the redox centers play in the kinetics of the electron-transfer events? Second, is there a link between the thermodynamic driving force of the electron-transfer event and the rate at which it occurs? The theoretical analysis of this problem is by no means simple. It has been — and still is — a major challenge for theoreticians to develop models that give satisfactory explanations for all relevant experimental observations and to predict with good accuracy the kinetic characteristics of electron transfer events.

Chemists often consider it self-evident that bimolecular reactions necessarily proceed through the formation of an encounter complex, where the distance between the reacting molecules is equal to the sum of the gas kinetic or crystallographic radii. Electron-transfer reactions are an exception to this rule since they can occur over a range that is significantly longer than the collisional distance. This behavior arises from a quantum mechanical effect frequently referred to as tunneling: due to small mass, which is equivalent to a relatively long wavelength, the electron can penetrate through energy barriers through which motion is forbidden for particles by classical mechanics.

In order to derive the probability for electron tunneling one must solve the time-independent Schrödinger equation:

$$H\Psi = E \cdot \Psi \tag{1}$$

It is instructive to consider the simple case where the energy barrier has a rectangular shape (Figure 1). The electron is considered to impinge with a total energy E on the left side of the barrier. Within the forbidden zone the potential energy increases to the value V. In the region of the barrier the Hamiltonian has the value

$$H = -\frac{\hbar^2}{2m_e}\frac{d^2}{dx^2} + V \tag{2}$$

and integration of Equation 1 using this operator gives the electron wave function

$$\Psi = Ae^{-\kappa x} + Be^{\kappa x} \tag{3}$$

where A and B are constants, $\kappa = \{2m_e(V - E)/\hbar^2\}^{1/2}$ corresponds to the wave vector of the electron in the classically forbidden zone multiplied by $\sqrt{-1}$, and (V − E) corresponds to the barrier height. In Equation 3 the first term reflects the movement of the electron from the left to the right, while the second exponential expresses the motion in the inverse sense. For infinitely wide barriers, the second term is zero and the amplitude of the wave function as well as the probability density $|\psi|^2$ for finding the electron in the classically forbidden zone decrease exponentially with distance:

$$\Psi = Ae^{-\kappa x} \tag{4}$$

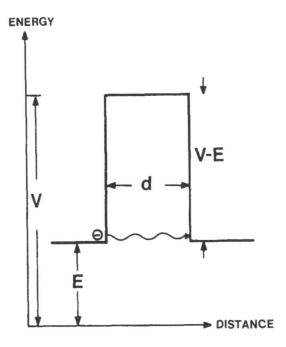

FIGURE 1. Schematic illustration of electron tunneling through a barrier of width d. E is the total energy of the electron impinging on the barrier from the left. V is the potential energy at the top of the barrier. Within the barrier the electron has negative kinetic energy.

In Equation 4 the wave vector κ may be interpreted as the reciprocal of the distance over which the wave function has declined to $1/e$ of the initial value. It has a value of 0.52 Å^{-1} for a barrier height of 1 eV.

For an energy barrier of finite width d it is of interest to calculate the tunneling probability. This is the probability that an electron impinging on the barrier on the left will emerge from it on the right side. It can be readily calculated from the ratio of the square of the wave function amplitudes before and after the barrier.[2] The expression obtained is

$$P = 1/[1 + f(d, E, V)] \qquad (5)$$

where

$$f(d, E, V) = \frac{(e^{\kappa d} - e^{-\kappa d})^2}{16(E/V)(1 - E/V)} \qquad (6)$$

In Figure 2 we give a graphical illustration of the tunneling probability of electron across a 5-Å wide barrier as a function of initial energy E for V values of 2 and 1 eV. Classical physics impose the condition $E \geq V$ in order for the electron to surmount the barrier. Quantum mechanical tunneling, on the other hand, allows the penetration of the barrier by electrons whose initial energy is much below V. This is the origin of long-range electron transfer which is very important for redox reactions in chemical and biological systems.

Using these quantum mechanical concepts, Gamov[3] derived a simple expression for the average transition time required for a particle with energy E to cross a potential barrier whose shape is described by the function V(x):

$$\tau^{-1} = V_0 \exp(-\beta d) \qquad (7)$$

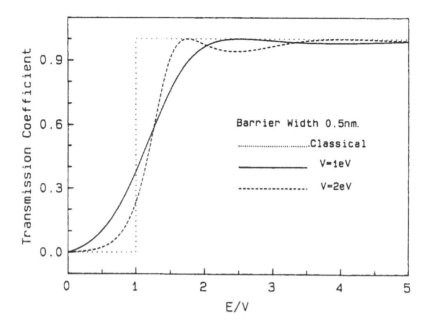

FIGURE 2. Plot of the transmission factor P for electron tunneling across a barrier of 5-Å width as a function of the ratio E/V. P was calculated from Equations 5 and 6 for potential energies of 1 and 2 eV.

where V_0 is a frequency factor, d is the width of the barrier, and β is the damping coefficient:

$$\beta = \frac{(8m_e)^{1/2}}{\hbar} \int_0^d (V(x) - E)^{1/2} \, dx \qquad (8)$$

Note that β depends on the shape and height of the barrier, which frequently is difficult to calculate. For a rectangular barrier it is related to the wave vector via $\beta = 2\kappa$. Strictly speaking, Gamov's formula is only valid for isolated particles in free space whose energy level distribution is continuous. Therefore, it cannot be applied to quantitative kinetic analysis of homogeneous and heterogeneous charge transfer processes involving donor and acceptor sites where the electron wave function is localized and the energy states are quantized. Nevertheless, Equation 8 has been useful[4] to rationalize in a qualitative manner some characteristic features of long-range electron-transfer processes such as the exponential dependence of the reaction rate on the distance between donor and acceptor.

In modern electron-transfer theory one distinguishes between adiabatic and nonadiabatic reactions. Adiabatic electron-transfer reactions are characterized by a relatively strong electronic interaction between the reactants. They can be treated by a classical transition state theory. Central to this approach is the idea that the reaction proceeds through a fairly well-defined activated complex (Figure 3). In the initial phase of the reaction a precursor complex is formed between the electron donor and acceptor. Electron transfer proceeds subsequently along the reaction coordinate from the reactant valley via the transition state to the successor complex:

$$(A...D) \xrightarrow{k_{et}} (A^-...D^+) \qquad (9)$$

The rate constant k_{et} is related to the free energy of activation ΔG^* via

$$k_{et} = v_0 \exp(-\Delta G^*/RT) \qquad (10)$$

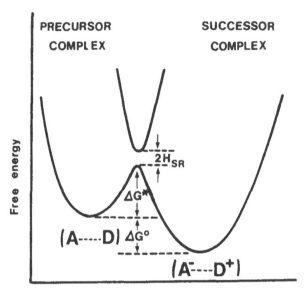

FIGURE 3. Free energy diagram for an adiabatic electron transfer reaction from a donor D to an acceptor A. A- - -D indicates the nuclear configuration of the precursor complex, while $A^- - - -D^+$ refers to that of the successor complex. ΔG^0 and ΔG^* are the standard free energies of reaction and activation, respectively.

where v_0 is the frequency of nuclear reorganization giving rise to the formation of the activated complex. From the simple transition state theory $v_0 = kT/h$ where k is the Boltzmann constant. Thus, the frequency parameter should be 6.6×10^{12} sec^{-1} at room temperature.

A more detailed model[6] expresses v_0 in terms of the frequencies of the vibrational modes and the free energy change associated with the formation of the activated complex. The formation of the activated complex is accompanied by changes in the nuclear configuration of the reactants (inner sphere organization) as well as a rearrangement of the solvation shell (outer sphere reorganization). The amounts of free energy λ required to bring about the inner and outer sphere reorganization processes are designated with a subscript in and out, respectively. Designating the frequencies of the vibrational modes involved in the activated complex formation with v_{in} and v_{out}, one obtains the frequency factor

$$v_0 = \frac{v_{out}^2 \lambda_{out} + v_{in}^2 \lambda_{in}}{4\Delta G^*} \qquad (11)$$

The nuclear coordinates of the activated complex correspond to a configuration where the free energies of the precursor and successor complex are identical, i.e. to the intersection of the two free energy curves in Figure 4 describing the precursor and successor state, respectively. Only in such a configuration can electron transfer occur without violation of the Franck-Condon principle. In the intersection region the potential surface splits into a lower and upper part, the separation being given by the electronic matrix element H_{SR} which couples the reactant and product state. The stronger the electronic interaction in the activated complex, the larger the splitting. The term adiabatic refers to the case where the system during electron transfer remains on the lower free energy surface. The criterion for adiabaticity is that the value of H_{SR} is at least 0.025 eV at room temperature. At smaller interaction

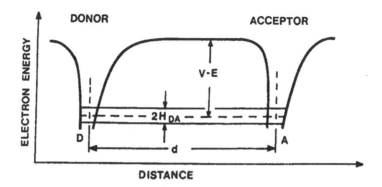

FIGURE 4. Nonadiabatic electron transfer from a donor D to an acceptor A separated by the distance d. $V - E$ is the height of the tunneling barrier, and H_{DA} the electronic coupling matrix element.

energies, crossing from the lower to the upper potential curve can occur, reducing the reaction probability and rendering the electron transfer nonadiabatic. In the adiabatic regime the electrons perfectly follow the nuclear motion. They have no important dynamics and do not undergo tunneling transitions independently of the displacement of the nuclei prescribed by the reaction coordinate. Therefore, electron tunneling, in particular distance dependence, is not considered in the adiabatic model. Nuclear tunneling, on the other hand, can occur and corresponds to a reaction channel where the system penetrates horizontally through the potential barrier in Figure 3. In such a case, electron transfer takes place without the reactants reaching the free energy of the activated complex. Nuclear tunneling plays a role only at low temperatures where the electron transfer is often faster than predicted by Equation 10. This leads to a positive deviation of the rate constants from the Arrhenius plot.

II. ADIABATIC OUTER SPHERE ELECTRON-TRANSFER REACTIONS IN SOLUTIONS: MARCUS THEORY

Adiabatic outer sphere electron-transfer reactions are frequently analyzed within the framework of the Marcus theory. Marcus[5] used a statistical mechanical model to describe the nuclear arrangements necessary for the formation of the activated complex. In his considerations he included the solution shell of the reactants as well as polarization effects in the solvent. The relation derived between the free energy of activation ΔG^* and the standard free energy of the electron-transfer process ΔG°_{et} is

$$\Delta G^* = \frac{(\Delta G^\circ_{et} + \lambda)^2}{4\lambda} \qquad (12)$$

Note that G°_{et} refers to the free energy change associated with the conversion of the precursor into the successor complex. The free energy of the overall electron-transfer reaction comprises the additional terms W_p and W_s which represent the work necessary to form the precursor state from the reactants and the successor state from the products, respectively:

$$\Delta G^\circ_{et} \text{ (overall)} = \Delta G^\circ_{et} + W_p - W_s \qquad (13)$$

Thus, the work terms contain free energy contributions from processes which are not directly associated with the electron transfer event. For an intramolecular electron transfer this could be a conformational change. For a bimolecular solution reaction W_p represents the work necessary to bring the acceptor and donor together. If electrostatic interactions make the main contributions, the Debye-Hückel theory gives

$$W_p = \frac{q_A \cdot q_A \cdot N_A}{\epsilon \cdot r(1 + \kappa r)} \qquad (14)$$

where N_A is Avogadros number, q_A and q_D are the charges of the acceptor of the solvent and donor, respectively; ϵ is the static dielectric constant; r is the distance between the centers of the acceptor and donor molecules in the precursor complex; and κ is the reciprocal Debye length defined as

$$\kappa = \left(\frac{8\pi N_A c^2 \mu}{1000\epsilon R}\right)^{1/2} \qquad (15)$$

where μ is the ionic strength and e is the charge on the electron.

The key parameter in Equation 12 is the reorganization energy. It is the free energy that would be required to move all the atoms constituting the precursor complex, including the solvation shell, from the equilibrium positions to the equilibrium positions of the successor complex without transferring the electron (Figure 3). The free energy of reorganization is usually separated into two terms:

$$\lambda = \lambda_{in} + \lambda_{out} \qquad (16)$$

where λ_{in} represents the inner sphere contribution arising from the activation of frequency modes within the precursor complex, and λ_{out} is the outer sphere term arising from the changes of the configuration of the solvent molecules. λ_{in} is obtained from the parameters of the inner sphere vibrational modes:

$$\lambda_{in} = 1/2 \sum_i k_i(d_s - d_p)_i^2 \qquad (17)$$

where d_s and d_p are the equilibrium bond distances of the ith vibrational mode in the successor and precursor complex, respectively, and k_i is the average force constant of this mode, which is related to the vibrational frequency via

$$k_i = 4\pi^2 v_i^2 \mu_i \qquad (18)$$

where μ_i is the reduced mass of the vibrating atoms. The value of k_i is derived from the force constants of the ith vibrational mode before (f_i^p) and after (f_i^s) electron transfer:

$$k_i = \frac{2f_i^s \cdot f_i^p}{f_i^s + f_i^p} \qquad (19)$$

The summation in Equation 17 extends over all vibrational modes of the reacting molecules.

The outer sphere reorganization energy is derived from the polarizability of the solvent considered to be a continuous polar medium:

$$\lambda_{out} = (\Delta e)^2 \left(\frac{1}{2r_D} + \frac{1}{2r_A} - \frac{1}{r}\right)\left(\frac{1}{D_{op}} - \frac{1}{D_s}\right) \qquad (20)$$

where Δe is the charge transferred from donor to acceptor, r_D and r_A are the radii of the reactants, r is the distance between the reactant centers in the transition state, D_{op} is the square of the refractive index of the medium, and D_s is the static dielectric constant.

From Equation 12 it is apparent that for a series of electron transfer reactions having the same value of λ and different values of the driving force $-\Delta G_{et}^\circ$ the rate should attain a maximum when $-\Delta G_{et}^\circ = \lambda$. In other words, the driving force is optimal when the value corresponds to the reorganization energy.

A case of special interest is that of a self-exchange reaction, e.g.,

$$Fe(H_2O)_6^{2+} + Fe(H_2O)_6^{3+} \rightarrow Fe(H_2O)_6^{3+} + Fe(H_2O)_6^{2+} \tag{21}$$

or

$$Ru(bipy)_3^{2+} + Ru(bipy)_3^{3+} \rightarrow Ru(bipy)_3^{3+} + Ru(bipy)_3^{2+} \tag{22}$$

Here, the electron is exchanged between the oxidized and reduced state of the same species, and therefore there is no net chemical change. Since the free energy of the reaction is zero, Equation 12 simplifies to

$$\Delta G^* = 1/4\lambda \tag{23}$$

For outer sphere electron-transfer reactions which are accompanied by a net chemical change, i.e.,

$$A + D \rightarrow A^- + D^+ \tag{24}$$

the reorganization parameter λ_{12} can be estimated from the value for the exchange reactions of two individual couples, i.e., A/A^- (λ_{11}), and D^+/D (λ_{22}) by using the Marcus cross relation:

$$\lambda_{12} = 1/2(\lambda_{11} + \lambda_{22}) \tag{25}$$

Thus, using Equation 25 one can write Equation 12 as

$$\Delta G_{12}^* = \frac{1}{8}(\lambda_{11} + \lambda_{22}) + 1/2\Delta G_{12}^\circ + (\Delta G_{12}^\circ)^2/2(\lambda_{11} + \lambda_{12}) \tag{26}$$

The exponential of this expression after division by $-RT$ gives the rate constant for the cross reaction divided by the frequency factor. Equation 26 allows one to express the rate constant of the cross reaction in terms of the rate constant for the individual exchange reactions, i.e.,

$$k_{11} = v_{0,11} \exp(-\lambda_{11}/RT) \tag{27}$$

and

$$k_{22} = v_{0,22} \exp(-\lambda_{22}/RT) \tag{28}$$

Assuming that the frequency factor for the cross reaction. v_0 is the geometric mean of the frequency factors for the self-exchange reactions,

$$v_0 = (v_{0,11} \times v_{0,22})^{1/2} \tag{29}$$

one obtains

$$k_{12} = (k_{11} k_{22} K_{12} f)^{1/2} \qquad (30)$$

where

$$K_{12} = \exp(-\Delta G^{\circ}_{et}/RT) \qquad (31)$$

is the equilibrium constant for the cross reaction which can be derived from the standard redox potentials of the acceptor $[E^{\circ}(A/A^-)]$ and donor $E^{\circ}(D^+/D)$:

$$\Delta G^{\circ}_{et} = E^{\circ}(A/A^-) - E^{\circ}(D^+/D) - W_p + W_s \qquad (32)$$

The factor f in Equation 30 is defined by

$$\log f = (\log K_{12})^2/(4 \log(k_{11} \cdot k_{22}/v_O^2) \qquad (33)$$

So far, our discussion has focused on the electron transfer within a pair of reactants, the precursor complex (A...D) giving rise to the formation of the successor complex (A$^-$...D$^+$), and we have been able to relate the first-order rate constant k_{et} of this process to the driving force of the reaction and the reorganization energy λ. The value of the latter was derived from the reorganization energies of the self-exchange reactions yielding Equation 30. This can be directly applied to the case of intramolecular electron-transfer processes or other redox reactions involving donor and acceptor pairs that are held at a fixed distance, such as electrochemical processes where the depolarizer is chemically attached to the surface. Here k_{et} is the experimentally observable variable. However, in many other cases, the diffusional approach of the reactants to form an encounter complex precedes the electron-transfer step. This process needs to be taken into account when comparing theoretical predictions for k_{et} with experimental results. A bimolecular electron-transfer reaction in solution involves as the first step the formation of the precursor complex:

$$A + D \underset{k_{-d}}{\overset{k_d}{\rightleftharpoons}} (A...D) \qquad (34)$$

which is followed by electron transfer within the solvent case:

$$(A...D)_p \xrightarrow{k_{et}} (A^-...D^+)_s \rightarrow A^- + D^+ \qquad (35)$$

Applying the steady-state approximation to the concentration of precursor and successor complex, one obtains for the observed second-order rate constant

$$\frac{1}{k_{obs}} = \frac{1}{K} \times \frac{1}{k_{et}} + \frac{1}{k_d} \qquad (36)$$

where K and k_d are the equilibrium constant and diffusion-controlled rate constant for encounter complex formation, respectively. If the preequilibrium Equation 34 is always established during electron transfer, Equation 36 simplifies to

$$k_{obs} = Kk_{et} \qquad (37)$$

The equilibrium constant K has been interpreted in different ways, and these models have been discussed previously.[7] A recent analysis[15] relates K to the reaction layer thickness dR:

$$K = 4\Pi R^2 \, dR \qquad (38)$$

where $R = r_A + r_D$. A good estimate for dR is the reciprocal value of the tunneling damping factor $1/\beta$ (see Equation 7). Using $\beta = 1.2 \, \text{Å}^{-1}$ and a reaction radius R of 14 Å one obtains $K = 2 \times 10^{-21} \, \text{cm}^3$ per molecule, or $1.2 \, \text{M}^{-1}$. If the reactants are charged, one has to take the coulombic interaction energy into account, yielding

$$K' = 4\Pi R^2 dR \exp(-w_r kT) = K \exp(-w_r/kT) \qquad (39)$$

This formalism can be used to derive numerical estimates for the preexponential factor observed for outer sphere electron-transfer reactions in solution. Expressing the experimentally determined rate constant as

$$k_{obs} = A \exp(-w_r/kT) \exp(-\Delta G^*/kT) \qquad (40)$$

and assuming that the preequilibrium model can be applied, one obtains for adiabatic reactions

$$A = v_0 K \qquad (41)$$

Since the value of v_O is between 10^{11} and $10^{14} \, \text{sec}^{-1}$ and that of K is typically of the order of 0.1 to 1 M^{-1}, A is expected to lie within the range of 10^{10} to $10^{14} \, M^{-1} \, \text{sec}^{-1}$. By comparison, the diffusion-controlled rate constant for the precursor complex formation k_d, according to the Smoluchowski equation,

$$k_d = 4\Pi R(D_A + D_D) \qquad (42)$$

where D is the diffusion coefficient, is in the range of 10^9 to $10^{10} \, M^{-1} \, \text{sec}^{-1}$. Hence, A could be significantly larger than k_d. These considerations cast doubts on the frequently made assumption that A is equal to the rate of encounter complex formation. It should also be noted that in cases where $K \times k_{et} > k_d$ the preequilibrium assumption is no longer valid. Therefore, instead of Equation 37 the steady-state approximation, i.e., Equation 36, must be used in the evaluation of the experimental results.

As an example for the application of the Marcus theory to solution electron-transfer reactions, let us consider the exchange reactions (Equations 21 and 22). The $Fe(H_2O)_6^{3+}$-$Fe(H_2O)_6^{2+}$ electron exchange is associated with a significant decrease in the Fe-water bond distance ($d_2 - d_1 = 0.14$ Å) leading to a large inner sphere reorganization energy, $\lambda_{in} \simeq$ 1.5 eV.[7] By contrast, in the case of the $Ru(bipy)_3^{2+}$-$Ru(bipy)_3^{3+}$ couple, the bond distances are practically identical in the two oxidation states; hence, $\lambda_{in} \simeq 0$. Since during the $Fe(H_2O)_6^{2+}$-$Fe(H_2O)_6^{3+}$ exchange the two Fe centers approach each other more closely (R = 7 Å) than the Ru atoms during the $Ru(bipy)_3^{2+}$-$Ru(bipy)_3^{3+}$ exchange (R = 14 Å), the outer sphere reorganization energy is also higher, i.e., $\lambda_{out} = 1.1$ eV for the former as compared to 0.55 eV for the latter redox couple. Thus, the total reorganization energy for $Fe(H_2O)_6^{3+}$-$Fe(H_2O)_6^{2+}$ is predicted to be 2.6 eV, while that for $Ru(bipy)_3^{3+}$-$Ru(bipy)_3^{2+}$ is only 0.55 eV. Using a frequency factor of $10^{13} \, \text{sec}^{-1}$ for both couples, one expects for the electron-transfer rate constants at room temperature associated with Reactions 21 and 22 values of 100 and $5 \times 10^{10} \, \text{sec}^{-1}$ respectively.

These are monomolecular rate constants which refer to the conversion of the precursor into the successor complex. Assuming that the preequilibrium condition applies, the second-order rate constants are obtained by multiplying the k_{et} with K'. Using Equation 39 for K' and Equation 14 to calculate the electrostatic work W_p, one obtains for low ionic strength conditions where $\beta << r$, $K' = 5.2 \times 10^{-2}$ for the Ru(bipy)$_3^{2+}$-Ru(bipy)$_3^{3+}$ and $K' = 8.5 \times 10^{-4}$ for the Fe (H$_2$O)$_6^{3+}$-Fe(H$_2$O)$_6^{2+}$ exchange reaction. Thus, the bimolecular rate constant for Reaction 22 is predicted to be 2.6×10^9 M^{-1} sec^{-1}, which must be compared with the experimental estimate of 5×10^8 M^{-1} sec^{-1}.[7] For Reaction 21 the bimolecular rate constant at low ionic strength is predicted to be 0.1 M^{-1} sec^{-1} compared to 5 M^{-1} sec^{-1} obtained experimentally at 1 *M* ionic strength. The cross reaction between Fe(H$_2$O)$_6^{2+}$ and Ru(bipy)$_3^{3+}$ is predicted to have a rate constant of 5×10^6 M^{-1} sec^{-1} using Equations 12, 14, 25, and 40 with $A = 10^{12}$ M^{-1} sec^{-1}. Experimental values are between 1- to 5×10^6 M^{-1} sec^{-1}. Given the approximations involved in the theoretical estimate, the agreement is surprisingly good.

III. NONADIABATIC ELECTRON-TRANSFER REACTIONS

So far we have restricted ourselves to adiabatic electron-transfer reactions which were discussed within the framework of transition-state theory. The condition of adiabacity implies that during the conversion of the precursor into the successor complex the system remains on the same energy surface, i.e., curve 1 in Figure 3. Once the transition state has been reached the passage into the product valley is assumed to occur with 100% probability.

Many electron-transfer reactions in chemistry and biology are nonadiabatic. This is true in particular for heterogeneous electron-transfer events where the electron donor and acceptor are separated by an interface. In these cases the coupling between the reactant and product states can be relatively weak, rendering the model of adiabatic electron-transfer reactions discussed in the previous section inapplicable. The general expression for the electron-transfer rate constant, taking into account nonadiabacity, is

$$k_{et} = \frac{2\,\Pi}{\hbar}\,|H_{DA}|^2(FC) \tag{43}$$

Here, $|H_{DA}|^2$ is the square of the absolute value of the electron exchange matrix element which contains the distance dependence of the electron tunneling rate, while FC is the Franck-Condon factor expressing the contribution of nuclear reorganization to the overall probability of electron transfer. Equation 43 is based on the Born-Oppenheimer approximation which assumes that the motion of electrons and nuclei can be treated independently.

H_{DA} measures the interaction energy between the tails of the wave function of the electron on the donor and on the acceptor site ($\psi_D H_{DA} \psi_A d\tau$).

This exchange interaction leads to a splitting of the electron orbitals into two new mixed donor-acceptor orbitals. The energy gap between these two states is equal to $2H_{DA}$ (Figure 4). The higher the interaction energy, the faster the tunneling transition of the electron from the donor to the acceptor site. A very important feature is the distance dependence of H_{DA}. In general, H_{DA} falls exponentially with d beyond the distance of closest approach of acceptor and donor (contact distance $d°$):

$$|H_{DA}|^2 = |H_{DA}|^2\,(d = d°)\,\exp[-\beta(d - d°)] \tag{44}$$

Note that β is identical with the damping coefficient in Gamov's formula (Equation 7). The value depends on the shape and height of the tunneling barrier. Assuming a rectangular barrier of 2 eV height, one obtains from Equation 8 $\beta = 1.5$ Å^{-1}. Other theoretically estimated and experimentally inferred values of β range from 2.6 to 1 Å^{-1}. The value of 2.6 refers to a theoretical calculation where the electron tunnels from one reactant to the other via a vacuum.[9] For example, Beitz and Miller [10] determined from the distance dependence of the electron-transfer rates in frozen media $\beta = 1.2\,\text{Å}^{-1}$ while Hupp and Weaver[11] and Li and Weaver[12] obtained $1.3\,\text{Å}^{-1}$ for a heterogeneous charge transfer from a Au electrode to $Co(NH_3)_5^{3+}$ groups attached to the surface via chemical spacers. For electron tunneling through an array of fatty acid monolayers the damping factor is around $1\,\text{Å}^{-1}$.[13] Very recently, the distance dependency of the self-exchange rate of a homologue series of manganese (II) isonitrile complexes was analyzed and yielded a damping factor of $2.4\,\text{Å}^{-1}$.[14] Thus, there is by now experimental confirmation that the rate constant for electron tunneling in nonadiabatic reactions decreases exponentially with distance, as expected from the Gamov equation. Most of the β values determined so far are between the limits of 1 to $3\,\text{Å}^{-1}$. It should be noted at this point that the outer sphere reorganization energy is also distance dependent, Equation 20. This can complicate the experimental determination of β[6].

Theoretical predictions of the damping coefficient are difficult to make since this requires the knowledge of the exact shape of the energy barrier, which is seldom available. According to Equation 8 β should be proportioned to the square root of the barrier height integrated over the tunneling distance. This dependency was recently checked by the kinetic analysis of intramolecular electron transfer processes involving a donor and acceptor molecule separated by a rigid molecular spacer.[15] By judicious selection of the donor and acceptor species the barrier for electron tunneling could be changed without affecting the driving force or distance of electron transfer. Instead of the square root relation predicted by Equation 8 β was found to be proportional to the logarithm of the barrier height. This would indicate the presence of superexchange-type electronic interactions[16] involving indirect coupling of donor and acceptor orbitals through mixing with orbitals of the spacer groups.

Having dealt with the electron exchange matrix element and the distance dependency of electron-transfer rates, we shall now turn to the discussion of the FC factor in Equation 43. The latter expresses the effect of thermal activation of nuclear vibrations on the electron transfer probability. In the evaluation of FC one must distinguish the classical Marcus approach based on the concepts of transition-state theory, i.e., the notion of activated complex formation, from quantum mechanical models.

Quantum mechanical theories of electron-transfer reactions do not employ the transition-state theory. In other words, the formation of an activated complex is not considered. Thus, the intersection region of the two free energy curves in Figure 3 is deemphasized. The electron-transfer process is treated as a radiationless transition of the system from the reactant (precursor) to the product (successor) state. Fermi's "golden rule" is applied to derive the transition probability. The evaluation of FC is done either by the semiclassical or the full quantum mechanical method. The former considers the overlap of the two energy distribution functions for electronic states associated with the acceptor and donor system, while in the latter approach the spatial overlap of the vibrational wave functions of the reactant and product system is calculated.

Both the semiclassical and quantum mechanical model take the quantization of vibrational energy states into account, while the classical theory assumes a continuum of states. Thus, the mean energy of a classical harmonic oscillator is kT, whereas it is

$$u = \tfrac{1}{2}h\nu \coth(h\nu/2kT) \qquad (45)$$

for the quantized oscillator. At very low temperature the expression in Equation 45 has the value of $u = \tfrac{1}{2}\,h\nu$, while at a high temperature, where $h\nu < kT$, it approaches kT. Therefore, within the high-temperature limit the quantized and classical oscillator have the same mean energy. For room temperature conditions only low-frequency oscillators have $u = kT$, since, in order to attain this limit, the wave number for the vibrational transition is required to be smaller than $200\,\text{cm}^{-1}$. While this condition may be fulfilled for the low-frequency solvent modes involved in outer sphere reorganization, it does not apply to inner sphere vibrations for which $\tilde{\nu}$ values are typically far above this limit. It is evident that at a low temperature the classical oscillator model fails for all vibrational modes, including the low-frequency ones.

Having stated the principal differences among the classical, semiclassical, and quantum mechanical concepts we shall now elaborate in more detail on these three models of nonadiabatic electron transfer processes.

A. Classical Description of Nonadiabatic Electron-Transfer Reactions

Nonadiabacity is accounted for within the framework of the Marcus theory by introducing a correction factor, the electronic transmission coefficient κ, into Equation 10. Expressing the free energy of activation by Equation 12 one obtains

$$k_{et} = v_0 \kappa \exp[-(\Delta G^{\circ} + \lambda)^2/4\lambda RT] \tag{46}$$

where κ corresponds to the probability of electron transfer when the reactants are in the nuclear configuration of the transition state. For adiabatic reactions, $\kappa = 1$. In the nonadiabatic case the separation of the energy surfaces 1 and 2 in Figure 3 is small enough to permit oscillations from one surface to the other once the activated complex has been formed. As a consequence, the reaction probability is smaller than 1. The value of the transmission coefficient is derived by applying a quantum mechanical formalism based on Landau.[17] The probability for electron transfer in the transition state is expressed in terms of the electron exchange matrix element H_{DA}, which is twice the distance separating the lower from the upper energy surfaces in Figure 3;

$$v_0 \kappa = \frac{2\pi}{\hbar} |H_{DA}|^2 (4\pi\lambda kT)^{-1/2} \tag{47}$$

Combining Equations 46 and 47 gives the following rate constant for nonadiabatic electron transfer:

$$k_{et} = \frac{2\Pi}{\hbar} |H_{DA}|^2 (4\Pi\lambda kT)^{-1/2} \exp[-(\Delta G + \lambda)^2/4\lambda RT] \tag{48}$$

Comparison of Equation 48 with the general expression for nonadiabatic electron-transfer reactions given by Equation 42 identifies FC derived by the classical theory as

$$(FC) = (4\Pi\lambda kT)^{-1/2} \exp[-(\Delta G^{\circ} + \lambda)^2/4\lambda kT] \tag{49}$$

Using Equation 44 to express the distance dependency of the electron exchange matrix element H_{DA}, one obtains finally

$$k_{et} = \left(\frac{\Pi}{\lambda kT}\right)^{1/2} \frac{1}{h} |H_{DA}|^2 (d = d^{\circ}) \exp[-\beta(d - d^{\circ})] \exp\left(-\frac{(\Delta 6^{\circ} + \lambda)^2}{4\lambda kT}\right) \tag{50}$$

where $|H_{DA}|^2$ $(d = d^{\circ})$ is the value of the square of the exchange matrix element at the distance of closest approach between electron donor and acceptor.

It is sometimes assumed that the reaction is adiabatic when the donor and acceptor couple are at the distance of closest approach, becoming nonadiabatic only at larger separation (d

13

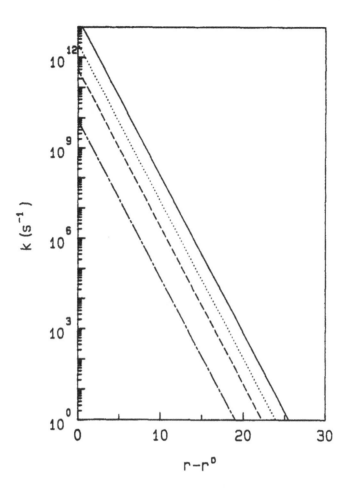

FIGURE 5.   Distance dependence of the electron transfer rate constant k_{et} according to Equation 48 for $v_0 = 10^{13}$ sec^{-1} and $\beta = 1.2$ Å^{-1}. ——, optimal driving force, $\Delta G^0 = -\lambda$; ········· $\lambda = 0.2$ ev, $\Delta G^0 = 0$; - - - -, $\lambda = 0.4$ eV, $\Delta G^0 = 0$; — · — · —, $\lambda = 0.8$ eV, $\Delta G^0 = 0$.

$> d^\circ$). In such a case the comparison of Equation 50 with 46 identifies the preexponential factor v_0 with

$$v_0 = \left(\frac{\Pi}{\lambda kT}\right)^{1/2} \frac{1}{h} |H_{DA}|^2 (d = d^\circ) \tag{51}$$

while the transmission coefficient is simply given by

$$\kappa = \exp[-\beta(d - d^\circ)] \tag{52}$$

Using $v_0 = 10^{13}$ sec^{-1} we have plotted in Figure 5 the distance dependence of k_{et} for several values of the free energy of the electron-transfer reaction. A damping factor of 1.2 Å was assumed, which is near the lower limit of experimentally measured values. It is readily seen that the maximal distance over which an electron can tunnel, say in 1 sec, is only 25 Å if the driving force for the charge transfer is optimal ($-\Delta G^\circ = \lambda$.) Reactions with zero driving force give smaller tunneling distances which decrease with increasing λ. Most of the photoinduced redox reactions which we shall consider in the following sections

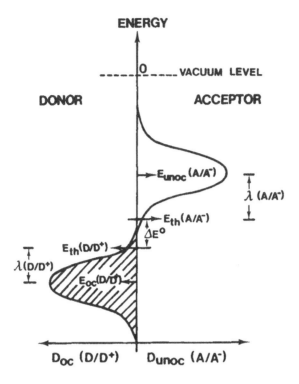

FIGURE 6.   Semiclassical picture of an electron transfer reduction from a donor D to an acceptor A, showing energy
distribution functions for occupied electronic states of the donor
and unoccupied electronic states of the acceptor system. For
clarity of presentation we have chosen a reaction with a positive
internal energy change resulting in a very small overlap of the
two distribution functions. The reorganization energies for the
individual redox couples are also indicated.

are in the 10^{-6} to 10^{-11}-sec time domain. These limits are imposed by the short lifetime of
electronically excited molecules or electron hole pairs in semiconductors. From Figure 5
one expects that during this time period electron tunneling over 5 and 10 Å may occur under
optimum conditions.

B. Semiclassical Description of Nonadiabatic Electron-Transfer Reactions

The semiclassical theory for nonadiabatic electron transfer is based on a model developed
by Förster[18] and Dexter[19] for the description of nonradiative transfer of electronic excitation
energy. The donor and acceptor are considered as two independent electronic subsystems.
Electron transfer involves tunneling from an occupied level of the donor into an unoccupied
state of the acceptor system (Figure 6). During the charge transfer the nuclear coordinates
and the total energy of the system are conserved. Therefore, the transition involves occupied
donor states that are isoenergetic with empty acceptor levels. The FC factor in Equation 43
is given by the integral of the energy distribution function $D_{DA}(E)$ of these isoenergetic sets
of states:

$$(FC) = \int D_{DA}(E) \, dE \tag{53}$$

The distribution function $D_{DA}(E)$ corresponds to the probability $P_{DA}(E)$ that an occupied
donor and unoccupied acceptor state occur within the same energy interval, i.e., between
E and E + dE, divided by the interval dE:

$$D_{DA}(E) = P_{DA}(E)/dE \tag{54}$$

Since the donor and acceptor redox systems are considered to be independent, $P_{DA}(E)$ can be written as the product of the individual probabilities for the occurrence of occupied donor and unoccupied acceptor levels, i.e.,

$$P_{DA}(E) = P_{oc}(E) \times P_{unoc}(E) \tag{55}$$

Using Equations 54 and 55, and expressing the probabilities in terms of the energy distribution functions for occupied and unoccupied states.

$$P_{oc}(E) = D_{oc}(D/D^+) \, dE \tag{56}$$

$$P_{unoc}(E) = D_{unoc}(A/A^-) \, dE \tag{57}$$

gives

$$FC = \int_{-\infty}^{+\infty} D_{oc}(D/D^+) \times D_{unoc}(A/A^-) \, dE \tag{58}$$

Hopefield[20] assumed the distribution functions $D_{oc}(D/D^+)$ and $D_{unoc}(A/A^-)$ to be Gaussian:

$$D_{oc} = C_1 \exp[-(E - E_{oc}(D/D^+)]^2/2\sigma^2) \tag{59}$$

$$D_{unoc} = C_2 \exp[-(E - E_{unoc}(A/A^-)]^2/2\sigma^2) \tag{60}$$

These functions have a maximum at $E_{oc}(D/D^+)$ for the occupied states of the donor and at $E_{unoc}(A/A^-)$ for the unoccupied states of the acceptor.

$E_{oc}(D/D^+)$ is the energy required for the vertical removal of the electron from the donor into the vacuum. The value is higher than the thermodynamic ionization energy of the donor $E_{th}(D/D^+)$ by an amount of $\lambda(D/D^+)$:

$$E_{oc}(D/D^+) = E_{th}(D/D^+) + \lambda(D/D^+) \tag{61}$$

$\lambda(D/D^+)$ corresponds to one half of the value of the reorganization energy for the exchange reaction:

$$D + D^+ \rightarrow D^+ + D \tag{62}$$

(It should be noted that the symbol λ was previously used for free energies of reorganization. Quantum mechanical calculations give energies, not free energies. The difference between these two variables is small when there is no significant entropy change associated with the reorganization.)

$E_{unoc}(A/A^-)$ is the energy gained by vertical attachment of the electron from the vacuum to the acceptor molecule. The value is lower than the thermodynamic electron affinity of A, $E_{th}(A/A^-)$, by an amount of $\lambda(A/A^-)$:

$$E_{unoc}(A/A^-) = E_{th}(A/A^-) - \lambda(A/A^-) \tag{63}$$

where the value of $\lambda(A/A^-)$ corresponds to one half of the reorganization energy for the exchange reaction:

$$A + A^- \rightarrow A^- + A \tag{64}$$

The parameter σ in Equations 59 and 60 corresponds to the width of the Gaussian distribution. Normalization — the integral of the distribution functions over all occupied donor and unoccupied acceptor state is 1 — shows the constant in these equations to have the value

$$C_1 = (2\Pi\sigma_D)^{-1/2} \qquad (65)$$

and   (66)

$$C_2 = (2\Pi\sigma_A)^{-1/2}$$

Inserting these distribution functions in Equation 58 and integration over all accessible energy states gives

$$(FC) = (2\Pi\sigma^2)^{-1/2} \exp[-(\Delta E^\circ + \lambda^2/2\sigma^2)] \qquad (67)$$

where in analogy to the Marcus cross relation it is assumed that $\lambda = \lambda(D/D^+) + \lambda(A/A^-)$ and that $\sigma^2 = \sigma_D^2 + \sigma_A^2$. Using the harmonic oscillator model to describe the reorganization of nuclear coordinates, one obtains for the width of the distribution functions

$$\sigma_D^2 = 2\lambda(D/D^+) \times u(D/D^+) \qquad (68)$$

and

$$\sigma_A^2 = 2\lambda(A/A^-) \times u(A/A^-) \qquad (69)$$

Here, $u(D/D^+)$ and $u(A/A^-)$ are the mean energies of the harmonic oscillators involved in the reorganization, of the donor and acceptor nuclear coordinates, respectively. In the semiclassical model developed by Hopefield,[20] these energies are expressed by Equation 45 which takes quantization of the vibrational levels into account. If only one frequency is involved in the reorganization $u(D/D^+) = u(A/A^-)$. Assuming, furthermore, that $\lambda(D/D^+) = \lambda(A/A^-)$, one obtains for FC the expression

$$(FC) = (2\Pi h\nu\lambda \coth(h\nu/2kT))^{-1/2} \exp\left[-\left(\frac{(\Delta E^\circ + \lambda)^2}{2h\nu \coth(h\nu/2kT)}\right)\right] \qquad (70)$$

which within the high temperature limit $(u = kT)$ reduces to

$$(FC) = (4\Pi\lambda kT)^{-1/2} \exp(-(\Delta E^\circ + \lambda)^2/4\lambda kT) \qquad (71)$$

Note that for electron-transfer processes with small reaction entropy Equation 71 becomes identical to the classical expression of the FC (Equation 49). It is interesting that the two approaches, which are based on entirely different concepts, converge under certain conditions to the same result. Inserting Equation 70 in Equation 43 gives for the electron-transfer rate constant

$$k_{et} = a_1(\tanh(T_c/T))^{1/2} \exp[a_2 \tanh(T_c/T)] \qquad (72)$$

where we have used the abbreviations

$$a_1 = (2\Pi/\hbar)|H_{DA}|^2(2\Pi h\nu\lambda)^{-1/2} \qquad (73)$$

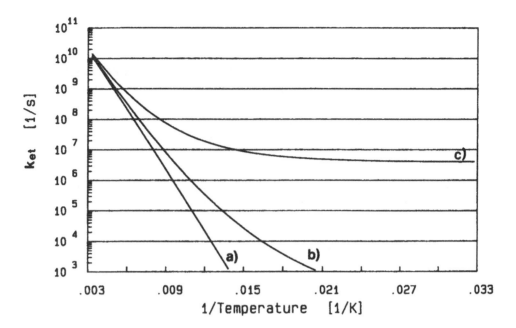

FIGURE 7. Comparison of the classical, semiclassical, and fully quantum mechanical theories for electron transfer reactions, showing the effect of temperature on the rate constant of a self-exchange reaction with $\Delta E^\circ = 0$, $\lambda = 0.55$ eV, and $|H_{DA}| = 0.01$ eV. (a) Classical Marcus theory (Equation 48); (b) fully quantum mechanical theory (Equation 89), hv = 0.02 eV; and (c) Hopfield's semiclassical theory (Equation 72), hv = 0.02 eV.

$$T_c = hv/2k \tag{74}$$

and

$$a_2 = (\Delta E^\circ + \lambda)^2/(2hv\lambda) \tag{75}$$

From Equation 72 the temperature dependence of the electron-transfer rate for a given donor-acceptor system is described by only three independent constants: a_1, a_2, and T_c. At low temperatures the hyperbolic tangent approaches 1, and the rate constant for electron transfer becomes

$$k_{et} = a_1 \exp(-a_2) \tag{76}$$

which is temperature independent. At higher temperature k_{et} depends on T, the slope of the Arrhenius plot of Equation 72 being

$$\frac{d \ln k_{et}}{d(1/T)} = \frac{T_c}{\sinh(2T_c/T)} - \frac{2a_2 T_c}{1 + \cosh(2T_c/T)} \tag{77}$$

Figure 7 illustrates the temperature dependency of k_{et} predicted by Equation 72. For simplicity we have chosen a self-exchange reaction ($\Delta G^\circ \simeq \Delta E^\circ = 0$) with a reorganization energy $\lambda = 0.55$ eV as is observed, for example, for the $Ru(bipy)_3^{2+}$-$Ru(bipy)_3^{3+}$ redox couple. A value of 0.01 eV was chosen for the matrix element, which corresponds to relatively strong electronic coupling.

The predictions of Equation 72 have been compared with experimental results, in particular with electron tunneling rates in biological systems.[21] Approximate agreement was found. In particular, the invariance of the electron-transfer rates with temperature at low temperatures was verified experimentally.

C. Quantum Mechanical Treatment of Nonadiabatic Electron-Transfer Reactions

It was pointed out above that the semiclassical treatment of electron-transfer reactions assumes the donor and acceptor redox couples to be a separate and independent set of oscillators. This model becomes inadequate if the donor and acceptor are coherently coupled to the same vibrational modes of the surroundings. In such a case, the full quantum mechanical treatment needs to be applied.

The quantum mechanical calculation of the FC factor in Equation 43 has been attempted in many different ways which have been reviewed in a recent monograph by De Vault.[22] This field was pioneered by the work of Huang and Rhys[23] which gave a theoretical analysis of the shape of optical absorption bands for F centers in solids. The spectra were interpreted in terms of coupling of lattice vibrations to the optical excitation of electrons trapped in the F centers. In the derivation of their equation, Huang and Rhys assumed that the oscillators coupled to electron excitation had only one frequency. A similar model was applied by Dogonadze et al.[24] and Levich and Dogonadze[25] to chemical redox reactions and by Jortner[26] to electron-transfer reactions in a biological environment. The equations obtained by these authors are practically the same as those derived by Huang and Rhys. In the early 1950s, Kubo[27] and Lax[28] developed a more general model which allows one to analyze cases where a variety of oscillator frequencies are coupled to electron transfer. The model also takes into account changes in the equilibrium positions and frequencies. For a single frequency the formula reduces to that of Huang and Rhys.

The simple case where the reactants and products involved in the electron-transfer reaction can be presented by two oscillators — or two sets of oscillators — with the same frequency is illustrated in Figure 8. The equilibrium distance x_p° in the product vibrational mode is usually different from that in the reactant mode, x_R°, due to the effect of coulombic forces acting on the polar groups of the oscillator. The displacement in the equilibrium position is expressed in quantitative terms by the Franck-Condon displacement factor S which depends on the frequency of the oscillator and the reorganization energy:

$$S = \frac{\lambda}{h\nu} = \frac{\Pi}{\hbar} \, \nu\mu(x_p^\circ - x_R^\circ)^2 \tag{78}$$

During the electron-transfer event, the total energy is conserved, which implies that the vibrational quantum state of the reactant system must be the same energy as that of the products. Thus, if the initial quantum number of the reactant oscillator is n and that of the product oscillator is n′,

$$n' = n + p \tag{79}$$

where pxhv $= \Delta E^\circ$ is the exothermicity of the reaction. This energy is dissipated into the medium surroundings of the product oscillator subsequent to electron transfer.

In the quantum mechanical model the nuclear factor (FC) is calculated from the overlap of isoenergetic vibrational wave functions of the reactant and product oscillator:

$$C(n, n') = \int \chi_n(R) \, \chi_{n'} \, (P) \, dx \tag{80}$$

For the harmonic oscillator, the wave function of a given vibrational level is the product of the Hermite polynomial:

$$H_n(x - x^\circ) = (-1)^n \exp[(x - x^\circ)^2](d^n/dx^n) \exp[-(x - x^\circ)^2] \tag{81}$$

REACTANTS | PRODUCTS

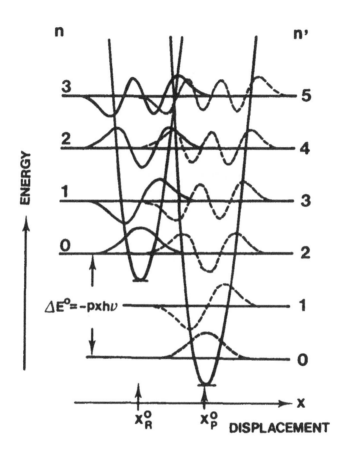

FIGURE 8. Quantum mechanical model for electron transfer reactions. The vibrations of the reactants and products are described by two harmonic oscillators with the same frequency. The equilibrium distance of the product oscillator is assumed to be larger than that of the reactants. The Franck-Condon displacement factor $S = \lambda/h\nu$ is used as a quantitative measure for the displacement of the equilibrium bond distance during electron transfer (see Equation 78). For the case considered the exothermicity of the reaction corresponds to two vibrational quanta ($p = 2$). The FC is calculated from the square of the overlap integral of the vibrational wave functions of the reactant and product oscillator in each vibrational level.

with $\exp[-(x - x^\circ)^2/2]$, where $(x - x^\circ)$ is the displacement of the oscillator from equilibrium position:

$$\chi_n = N_n H_n(x - x^\circ) \exp[-(x - x^\circ)^2/2] \tag{82}$$

The normalization constant N_n in Equation 82 has the value

$$N_n = \left[\left(\frac{2\nu\mu}{h}\right)^{1/2} \frac{1}{2^n n!}\right]^{1/2} \tag{83}$$

The square of the overlap integral in Equation 80, $C^2(n, n')$, corresponds to the strict definition of the FC factor. The value of the nuclear factor (FC) in Equation 43 is derived by multiplication of the FC factor of a given vibrational level n with the Boltzmann probability $P(n, T)$ that the reactant oscillator is excited to this level in the thermal equilibrium:

$$P(n, T) = \exp(-nh\nu/kT)[1 - \exp(-h\nu/kT)] \tag{84}$$

Summation of the product $P(n, T) \times C^2(n, n')$ for all vibrational states of the reactant oscillator and multiplication of the sum with the density of vibrational states, i.e., $1/h\nu$ yields

$$(FC) = \frac{1}{h\nu} \sum_{n=0}^{n=\infty} C^2(n, n') P(n, T) \tag{85}$$

Inserting this expression in Equation 43 gives for the rate constant of electron transfer

$$k_{et} = \frac{4\Pi^2}{h^2\nu} |H_{DA}|^2 \sum_{n=0}^{n=\infty} C^2(n, n') P(n, T) \tag{86}$$

If there are many modes with different frequencies involved in electron transfer, the summation of the FCs becomes very cumbersome. A suitable mathematical approach is the Kubo[27] and Lax[28], who employed generating functions to evaluate these sums. The expression obtained for the electron-transfer frequency k_{et} is

$$k_{et} = \frac{4\Pi^2}{h^2} |H_{DA}|^2 \int_{-\infty}^{+\infty} f(t) \exp(ih\Delta E^\circ t)\, dt \tag{87}$$

where

$$f(t) = \exp[-G + G_+(t) + G_-(t)]$$

$$G_+(t) = \sum_k S_k(\bar{n}_k + 1) \exp(i2\Pi\nu t)$$

$$G_-(t) = \sum_k S_k\bar{n}_k \exp(-i2\Pi\nu t)$$

$$G = G_+(t = 0) + G_-(t = 0)$$

The subscript k designates the different modes of the vibration in the multioscillator system, while $\bar{n}_k$ is the average vibrational quantum number in an ensemble of oscillators having the same frequency. The latter is obtained by Equation 45 from the mean energy of the oscillator:

$$\bar{n} = (u - h\nu/2)/h\nu = \exp(h\nu/kT - 1)^{-1} \tag{88}$$

In the case where there is only one single vibrational mode associated with electron transfer, Equation 87 reduces to

$$k_{et} = \frac{4\Pi^2}{\hbar^2\nu} |H_{DA}|^2 \exp[-S(2\bar{n} + 1)(\bar{n} + 1)/\bar{n}]^{p/2}\, xL_p[2S\bar{n}/(\bar{n} + 1)]^{1/2} \tag{89}$$

where $I_p(Z)$ is the modified Bessel function defined as

$$I_p(Z) = \sum_{j=0}^{\infty} \frac{(Z/2)^{p+2j}}{j!(p+j)!} \tag{90}$$

with $Z = 2S[\bar{n}/(\bar{n}+1)]^{1/2}$.

Equation 89 will be used in Chapter 3 to interpret the temperature dependence of electron-hole recombination kinetics in colloidal semiconductors.

At very low temperature, $\bar{n}$ approaches and Equation 89 becomes

$$k_{et}(T \to 0^\circ K) = \frac{4\Pi^2}{\hbar^2 v} |H_{DA}|^2 \exp(-S) \, S^p/p! \tag{91}$$

while in the high-temperature limit Equation 89 reduces to

$$k_{et} = \frac{2\Pi}{\hbar} |H_{DA}|^2 (4\Pi\lambda kT)^{-1/2} \exp\left[-\left(\frac{(\Delta E^\circ + \lambda)^2}{4\lambda kT}\right)\right] \tag{92}$$

which is the same as the classical expression, or the semiclassical one in the high-temperature approximation. Thus, the fully quantum mechanical and the semiclassical models reproduce, within the high-temperature limit, the Arrhenius activation behavior without the assumption that the system must pass over an activation barrier. In Figure 7 we compare the predictions of the three models using as an example a self-exchange reaction with $\lambda = 0.55$ eV, $hv = 0.02$ eV, and $H_{DA} = 0.1$ eV. The rate constants converge at higher temperatures. In the low-temperature range the classical model predicts too low values for k_{et} since it does not take into account nuclear tunneling.

IV. HETEROGENEOUS ELECTRON-TRANSFER REACTION

In this section, we shall discuss the salient kinetic features of heterogeneous electron-transfer processes and compare these to homogeneous reactions. Heterogeneous charge transfer events involve systems consisting of at least two different phases. For example, the electron transfer might occur between a metal or semiconductor electrode and a donor or acceptor in solution. In this case, the electron is delivered to or withdrawn from a solid, the electrode, which serves as an electron sink or source. Alternatively, the redox reaction could take place in a microheterogeneous system. The vast majority of biological electron-transfer reactions are microheterogeneous in nature. These comprise such fascinating processes as life–sustaining mitochondrial respiration and plant photosynthesis. Of great importance, in the context of the present monograph, are electron-transfer events in biomimetic devices, constituted, for example, by micelles or bilayer vesicles. Furthermore, redox reactions involving catalysis or photocatalysis by ultrafine spherical microelectrodes, e.g., colloidal metals and semiconductors, are of particular interest.

A common feature of all these systems is the presence of a phase boundary which in most cases is electrically charged. As a consequence, an electrical potential difference is established across the interphase which influences in a decisive fashion the charge transfer events. It is the control of the thermodynamics and kinetics of heterogeneous redox reactions by the electrical field present at the phase boundary which renders them unique with respect to the homogeneous analogues.

An additional feature which is inherent to microheterogeneous solution systems is the colloidal dimension of the dispersed phase. This leads to the sequestering of the species participating in the charge transfer reactions. Frequently, the situation is encountered where the redox events take place within ensembles of a few molecules, and there is no interaction between the ensembles present in different host aggregates, which remain isolated on the time scale of the reaction. It is apparent that in such a case the rate laws derived for homogeneous solutions fail to give an appropriate description of the reaction dynamics encountered with such systems. Therefore, new concepts have to be developed in order to account for the restricted size of the reaction space. These include statistical considerations concerning the distribution of reactants over the host aggregates. Furthermore, in a situation where the location of the donor or acceptor is restricted to the surface of the colloidal particle, the reduction in dimensionality of the reaction space has to be taken into account. These kinetical and statistical features are specific to microheterogeneous assemblies and shall be addressed in more detail in Chapters 2 and 3. In the following, we shall restrict ourselves to presenting a general description of the most important aspects of redox reactions in such multiphase systems.

A. Energetics of Interfacial Electron-Tranfer Reactions, Fermi Level, and Redox Potential of a Solution

Before addressing the problem of interfacial electron-transfer kinetics, it is profitable to briefly review the energetics of such redox processes. Consider a system where an electron is transferred from a donor to an acceptor located in two different solution phases, e.g., a membrane or micelle and water (Figure 9):

$$(D)_I + (A)_{II} \rightarrow (D^+)_I + (A^-)_{II} \tag{93}$$

Let Φ_I be the inner (Galvani) potential of phase I and Φ_{II} that of the phase II. The thermodynamic driving force for the electron transfer is then given by

$$\Delta\overline{G} = \sum v_i\overline{\mu}_i = \overline{\mu}_{D^+} + \overline{\mu}_{A^-} - \overline{\mu}_A - \overline{\mu}_D \tag{94}$$

where $\overline{\mu}_i$, the electrochemical potential of the species i, is defined as

$$\overline{\mu}_i = \mu_i + z_iF\phi \tag{95}$$

In Equation 95, μ_i is the chemical potential, z_i is the charge of the species, F is Faraday's constant, and ϕ is the electrical potential of the phase in which the species is located. Electrons will flow from the donor across the interface to the acceptor until equilibrium is established, where $\Delta\overline{G} = 0$. From Equation 94 the equilibrium condition is

$$\overline{\mu}_{D^+} - \overline{\mu}_D = \overline{\mu}_A - \overline{\mu}_{A^-} \tag{96}$$

and the potential difference between the two phases at equilibrium is

$$(\phi_{II} - \phi_I)_{eq} = \frac{1}{F}(\mu_{D^+} + \mu_{A^-} - \mu_D - \mu_A) \tag{97}$$

If the two phases behave as an ideal solution one obtains

$$(\phi_{II} - \phi_I)_{eq} = \frac{1}{F}(\mu^\circ_{D^+} + \mu^\circ_{A^-} - \mu^\circ_D - \mu^\circ_A) + RT \ln\left(\frac{C_{A^-} \cdot C_{D^+}}{C_A \cdot C_D}\right) \tag{98}$$

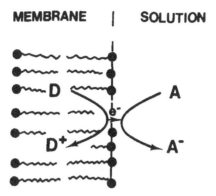

MEMBRANE | SOLUTION

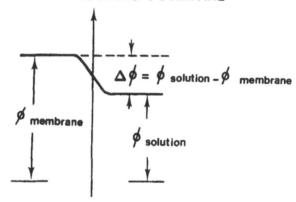

ELECTROSTATIC POTENTIAL

$$\Delta \phi = \phi_{\text{solution}} - \phi_{\text{membrane}}$$

ϕ_{membrane}

ϕ_{solution}

FIGURE 9. Schematic illustration of electron transfer from an electron donor located in a membrane (phase I, top) to an acceptor in the aqueous solution (phase II, bottom). The electrical potential difference between the two phases is $\Delta\phi$.

An alternative formulation of Equation 96 frequently employed is that the electrochemical potential of the electron in the two phases is equal:

$$(\bar{\mu}_{e^-})_I = (\bar{\mu}_{e^-})_{II} \tag{99}$$

This concept has been criticized on the grounds that in the solutions containing the redox couples there are no free electrons and that it is dubious to express an equilibrium condition in terms of a nonexisting species.[29] While this argument is certainly valid, there are justifications for using the Fermi level concept in order to describe the free energy of redox couples in solution. These arise from the consideration of electrochemical equilibria involving a solid conductor which is in contact with a solution containing the redox couple.[30] For example, if electron transfer occurs from a metal or semiconductor to an acceptor present in solution,

$$(ne^-)_{\text{solid}} + A \rightleftarrows A^{n-} \tag{100}$$

the equilibrium condition is

$$n(\bar{\mu}_{e^-})_{\text{solid}} = \bar{\mu}_{A^{n-}} - \bar{\mu}_A \tag{101}$$

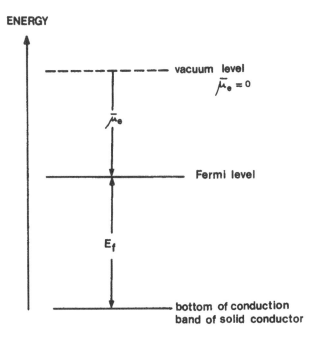

FIGURE 10.   Graphic interpretation of the kinetic Fermi energy and the electrochemical potential of an electron in a solid conductor.

where

$$(\bar{\mu}_e-)_{solid} = (\mu_e-)_{solid} - F\phi_{solid} \tag{102}$$

is the electrochemical potential of the electron in the solid. The physical significance of $\bar{\mu}_{e\bar{v}}$ is illustrated graphically in Figure 10. In the solid, the distribution of electrons over the available states follows Fermi statistics, the average occupancy of a level being given by

$$f(E) = (\exp[(E - E_f)/kT] + 1)^{-1} \tag{103}$$

The parameter E_f in Equation 103 is the kinetic Fermi energy of the electron. This energy corresponds to a state which on the average is occupied by 0.5 electrons. Note that E_f is expressed on an energy scale whose zero point is the bottom of the conduction band of the semiconductor or the metal. From a simple quantum mechanical model one obtains for T $= 0°K$:

$$E_f = (h^2/8\Pi^2 m_e)(3\Pi^2 n_e)^{2/3} \tag{104}$$

where m_e is the mass of the electron and n_e the number of conduction band electrons per unit volume of solid. E_f must be distinguished from the electrochemical potential $\bar{\mu}_e^-$ which corresponds to the free energy that is gained upon transferring the electron from a vacuum (where it is assumed to be at rest) into the Fermi level of the metal or semiconductor. Note that $\bar{\mu}_{e-}$, in contrast to E_f, is a function of the inner potential of the solid conductor.

As was pointed out above, the difference of $\bar{\mu}_{A-} - \bar{\mu}_A$ in Equation 101 is frequently referred to as the Fermi level of the redox electrolyte. Expressing the electrochemical potentials by Equation 95 gives

$$E_f(redox) = \mu_{A^{n-}} - \mu_A - nF\phi_{solution} \tag{105}$$

or, under standard conditions:

$$E_f^{\circ}(\text{redox}) = \mu_{A^{n-}}^{\circ} - \mu_A^{\circ} - nF\phi_{\text{solution}} \tag{106}$$

In the following, we establish a relation between E_f (redox) and the standard redox potential of the A/A^{n-} redox couple. It is a convention in electrochemistry to use the normal hydrogen electrode (NHE) as a reference. In this case, the redox potential E°_{NHE} (A/A^{n-}) reflects the Gibbs free energy $\Delta G^{\circ}_{\text{NHE}}$ of the solution reaction

$$\frac{n}{2} H_2 + A \rightarrow A^{n-} + nH^+ \tag{107}$$

under standard conditions, where

$$\Delta G^{\circ}_{\text{NHE}} = -nFE^{\circ}_{\text{NHE}} \tag{108}$$

Instead of expressing redox potentials with respect to the NHE, the vacuum or absolute potential scale is sometimes employed. In the latter case one considers the equilibrium

$$ne_{\text{vac}}^- + A \rightleftarrows A^{n-} \tag{109}$$

for which

$$\Delta G^{\circ}_{\text{vac}} = \mu_{A^{n-}}^{\circ} - \mu_A^{\circ} = -nFE^{\circ}_{\text{vac}} \tag{110}$$

assuming that the chemical potential of the electron at rest in vacuum at infinite distance from the electrode is zero. Comparison of Equations 110 and 106 yields finally

$$E_f^{\circ}(\text{redox}) = -nFE^{\circ}_{\text{vac}} - nF\phi_{\text{solution}} \tag{111}$$

which shows that the Fermi level of the electron in solution is not identical with the redox potential expressed on the vacuum scale. This has been a point of contention, particularly in the field of photoelectrochemistry, where it is frequently asserted[31] that these two quantities are identical. Khan and Bockris[32] have recently criticized this assumption. They pointed out that only in the unlikely case where the inner solution potential is zero does the condition E_f (redox) $= -nFE$ hold, as is evident from Equation 111.

Relating E°_{vac} to E°_{NHE} is straightforward since Equation 109 is obtained from Equation 107 by adding the reaction

$$ne_{\text{vac}}^- + (nH^+)_{\text{water}} \rightleftarrows \left(\frac{n}{2} H_2\right)_{\text{water}} \tag{112}$$

for which the standard free energy is known to be $-n$ (4.5 ± 0.3) eV.[33] Thus, in order to convert a redox potential measured against the normal hydrogen electrode (NHE) to that expressed on the vacuum scale one uses the relation

$$E^{\circ}_{\text{vac}} = E^{\circ}_{\text{NHE}} + 4.5 \tag{113}$$

For a detailed discussion of the conversion of relative to absolute potentials the reader is referred to a recent series of papers.[34-36]

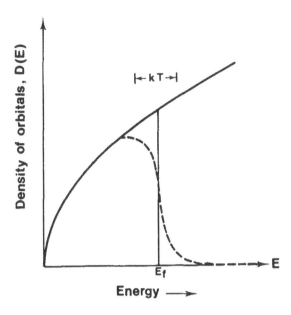

FIGURE 11. Density of electronic states as a function of energy for a free electron gas in three dimensions. The dashed curve represents the density of filled orbitals.

B. Distribution of Electronic Energy Levels in Solid Conductors and Electrolytes
1. Metals

A good approximation is to treat the electrons as a gas of noninteracting particles subject to the Pauli principle. Forces between the electron and the metal ion cores constituting the lattice of the solid are neglected. Thus, the electron is assumed to have only kinetic energy. In such a case, the density of electronic states (orbitals)

$$D(E) = \frac{dN(E)}{NdE} \tag{114}$$

is a parabolic function of energy[30] which is illustrated in Figure 11:

$$D(E) = \frac{V}{2\Pi^2 N} \left(\frac{2m_e}{h^2}\right)^{3/2} E^{1/2} \tag{115}$$

Here, N is the total number of states available in the volume V. The dashed line represents the distribution function of occupied orbitals obtained by multiplying D(E) by the Fermi factor given by Equation 103:

$$D_{oc}(E) = D(E) \times f(E) \tag{116}$$

The energy distribution function for empty orbitals, on the other hand, is simply

$$D_{unoc}(E) = 1 - D_{oc}(E) \tag{117}$$

at the Fermi energy, $D_{oc}(E) = D_{unoc}(E)$. An expression relating the Fermi energy at T = 0°K to the concentration of conduction electrons in the solid has already been given above, in Equation 104. This is obtained by integrating Equation 115 between the energy limits of

0 and E_f. At 0°K, each state up to the Fermi level contains one electron. Hence $N(E = 0$ to $E = E_f) = n_e \times V$, where V is the volume and n_e the number of electrons per unit volume.

Note that the value of the Fermi energy for most metals is between 2 and 6 eV, while the width of the energy domain where f(E) falls from 1 to 0 is of the order of kT, i.e., only 0.025 eV at room temperature. Therefore, it is reasonable to assume that in a metal the orbitals up to the Fermi energy are filled with electrons while those above E_f are empty.

2. Semiconductors

The most notable difference from metals is the appearance of a band gap, i.e., an energy domain with no accessible states for the electron. This is a consequence of the diffraction of the electron wave by the ion cores of the host lattice. Consider the simple example of a monoatomic linear lattice of lattice constant a. The Bragg condition for constructive interference is

$$\vec{k} = \pm n\Pi/a \tag{118}$$

where $\vec{k}$ is the wave vector of the electron and n is an integer. The constructive interference at $k = \pm \Pi/a$ arises because the wave reflected from one atom in the linear lattice interferes with phase difference 2π with the wave reflected from a nearest neighbor atom. The region in k space between $-1/a$ and $+1/a$ is the first Brillouin zone of this lattice. In Figure 12 we illustrate by means of an energy vs. wave vector diagram the appearance of a forbidden band as a consequence of Bragg reflection of the electron wave by the semiconductor lattice.

In a three-dimensional lattice the situation is more complex since the density of states depends on the direction in the crystal. In solid-state physics, it is customary to use also in this case an energy vs. wave vector representation, and we show a typical diagram for a direct band gap semiconductor in Figure 13A. The corresponding density of states function in Figure 13B is obtained by integrating over the entire k space. The analytical expression for the density of states function is

$$D(E) = \frac{V}{2\Pi^2 N} \left(\frac{2^* m_e}{h^2}\right)^{3/2} (E - E_{cb})^{1/2} \tag{119}$$

for conduction band levels and

$$D(E) = \frac{V}{2\Pi^2 N} \left(\frac{2^* m_h}{h^2}\right)^{3/2} (E_{vb} - E)^{1/2} \tag{120}$$

for valence band levels. In these equations, E_{cb} and E_{vb} correspond to the energy of the conduction and valence band edges, respectively, while $^*m_e$ and $^*m_h$ are the effective masses of the electron and the hole, respectively.

Equation 119 is equivalent to the expression derived for the density of electronic states in a metal except that the electron mass is replaced by an effective mass $^*m_e$. The latter is related to the second derivative of the electron energy with respect to the wave vector:

$$^*m_{e^-} = \hbar^2/(d^2E/d\vec{k}^2) \tag{121}$$

i.e., to the curvature of the density of state functions in Figures 9 and 10. A high curvature is typical for a material, such as GaAs, which is distinguished by a broad conduction band

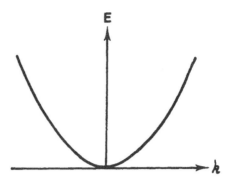

semiconductor

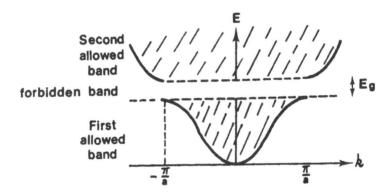

FIGURE 12. Energy vs. wave vector represetation of electronic states in a metal and a one-dimensional semiconductor. In the latter case, a forbidden band appears as a consequence of the Bragg diffraction of the electron wave by the lattice. The limits of the Brillouin zone are indicated.

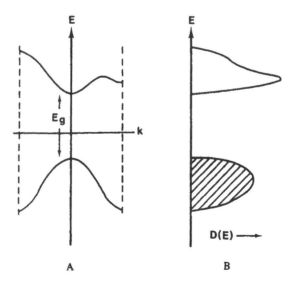

FIGURE 13. (A) Energy vs. wave vector representation of electronic states in a three-dimensional semiconductor with a direct band gap. (B) Density of electronic states D(E) as a function of energy obtained by integration over the whole space.

and a small effective electron mass, $*m_e = 0.07\ m_e$, resulting in a high electronic mobility. A small curvature, on the other hand, is found for semiconductors, such as TiO_2, whose conduction band is narrow and for which the electron has a high effective mass, $*m_e = 30\ m_e$, resulting in a low mobility. Note that the effective mass depends on electronic energy. The values quoted refer to thermalized electrons whose energy is close to the bottom of the conduction band. Often $*m_e$ is anisotropic depending on the direction of the electron motion within the semiconductor crystal.

Analogous considerations apply to the distribution function of valence band states (Equation 120) in which $*m_h$ is the effective mass of the hole.

As in the case of metals, the average occupancy of the energy levels by electrons and holes in semiconductors follows the Fermi-Dirac distribution given by Equation 103. In particular, for states near the bottom of the conduction band one obtains

$$c_{e^-}/n_{cb} = (\exp[(E_{cb} - E_f)/kT] + 1)^{-1} \tag{122}$$

where c_{e^-} is the concentration of conduction band electrons (the unit is number of charge carriers per (cubic centimeter), and n_{cb} is the number of states with energy close to the band edge in 1 cm³ of solid. The latter can be expressed in terms of the translational partition function of the conduction band electron, i.e.,

$$n_{cb} = 2z_{e^-}^{tr} = 2(2\Pi *m_e - kT/h^2)^{3/2} \tag{123}$$

Similarly, for the occupancy of valence band states close to the band edge one can write

$$c_{h^+}/n_{vb} = (\exp[(E_f - E_{vb})/kT] + 1)^{-1} \tag{124}$$

where the n_{vb}, the number of valence band levels in 1 cm³ solid, can be expressed by the translational partition function of the holes:

$$n_{vb} = 2z_{h^+}^{tr} = 2(2\Pi *m_h + kT/h^2)^{3/2} \tag{125}$$

Here, $*m_{h^+}$ is the effective mass of the valence band hole. If the energy difference between the Fermi level and the band edges is large compared to kT, Equations 122 and 124 can be written in the simplified form:

$$c_{e^-} = n_{cb} \exp[-(E_{cb} - E_f)/kT] \tag{126}$$

$$c_{h^+} = n_{vb} \exp[-(E_f - E_{vb}/kT] \tag{127}$$

Multiplication of these two equations gives the product of the electron and hole concentrations:

$$c_{e^-} \times c_{h^+} = n_{cb} \times n_{vb} \exp[-(E_{cb} - E_{vb})/kT] \tag{128}$$

which is a constant for a specific semiconductor at a given temperature. Furthermore, by taking the logarithm of Equations 126 and 127 and adding the two expressions one obtains

$$E_f = \frac{E_{cb} + E_{vb}}{2} + \frac{kT}{2} \ln\left(\frac{c_{e^-}}{c_{h^+}}\right) + \frac{3}{4} kT \ln\left(\frac{*m_{h^+}}{*m_{e^-}}\right) \tag{129}$$

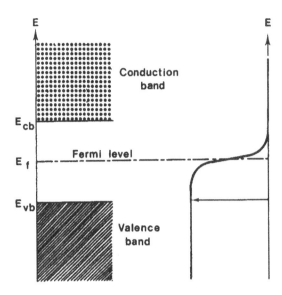

FIGURE 14.   Band picture and Fermi distribution function
for an intrinsic semiconductor.

This equation is very useful in that it relates the position of the Fermi level to the
concentration as well as the effective mass of electrons and holes. For intrinsic semiconductors, $c_{h^+} = c_{e^-}$ and the Fermi level is located approximately in the middle of the band
gap. This case is illustrated schematically in Figure 14. The effect of n and p doping is to
shift the Fermi level toward the conduction and valence band edges, respectively. The
knowledge of the position of the Fermi level, i.e., the electrochemical potential of the
electron, is important since it plays a decisive role in heterogeneous charge transfer reactions.

3. *Electronic Energy Levels in Solution*

The nature and distribution of electronic states associated with redox couples in solution
have been extensively discussed in the literature,[30,32] and only the most important aspects
will be reviewed here. The subject is best approached by considering the free energy changes
associated with electron removal from a donor or addition to an acceptor species. This has
already been briefly discussed in the previous section in connection with the semiclassical
theory of electron-transfer reactions. In the following we shall elaborate on this topic in
more detail.

Figure 15A illustrates the case where the electron is transferred from a donor in solution
to the vacuum reference level:

$$D \rightarrow D^+ + e_{vac} \tag{130}$$

Reaction 130 occurs in two consecutive steps. The first process involves the transition of
the electron from the solution into the vacuum. This is so fast that practically no displacement
of atoms is possible (vertical ionization). Therefore, the nuclear configuration of the oxidized
donor D^+ and the surrounding solvent molecules is at first the same as that of D at equilibrium.
In the second and slower step, the equilibrium configuration of D^+ and the solvation sphere
are formed. The free energy released during this relaxation process corresponds to $\lambda(D/D^+)$,
the energy of reorganization. Note that $\lambda(D/D^+)$ is one half of the value for the reorganization
energy of the (D/D^+) self-exchange reaction.

Consider next the reverse of reaction (Equation 72), i.e., the transfer of an electron from
the vacuum level to the oxidized donor in solution:

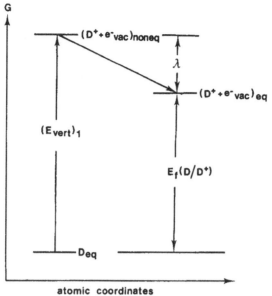

A

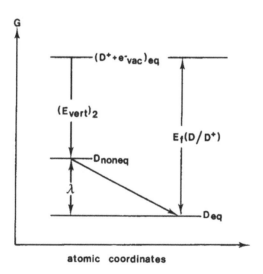

B

FIGURE 15.   (A) Free energy diagram for the removal of an electron from a donor in solution to a vacuum. The transition of the electron is so fast that a nonequilibrium state is first produced. Subsequent relaxation of vibrational modes of the donor and surrounding solvent molecules yields the equilibrium configuration. (B) Free energy diagram for the addition of an electron from a vacuum to an oxidized donor D^+ in solution. The subscripts noneq and eq refer to a nonequilibrium and an equilibrium configuration, respectively.

$$D^+ + e_{vac} \rightarrow D \tag{131}$$

Again, two steps can be distinguished: the electronic transition first produces D in a state with the nuclear coordinates (bond lengths, solvation state) of D^+. This is followed by the reorganization of the configuration, yielding the equilibrium configuration.

$(E_{vert})_1$ and $(E_{vert})_2$ can be considered as the energy of the occupied (E_{oc}) and unoccupied (E_{unoc}) electronic states of the D/D^+ redox system, respectively. The difference is equal to $2 \times \lambda(D/D^+)$:

$$E_{oc} - E_{unoc} = 2\lambda(D/D^+) \tag{132}$$

If there are N_D donor molecules present in solution there are also N_D occupied electronic states. Not all of these states are degenerate, i.e., at the same energy E_{oc}, since the energy of all the donor molecules is not identical. The energy of the donor molecules together with that of the neighboring solvent molecules is distributed over the different vibrational levels that are accessible to them at the temperature of the system. As a consequence, there is a whole set of occupied electronic levels. Similarly, in a solution containing N_D acceptor molecules there is a whole set of unoccupied electronic levels. Several models have been applied in order to derive the distribution function for the density of electronic states. It is commonly assumed that the different configurations arise from statistical fluctuations of the solvent molecules adjacent to D or D^+ about the equilibrium configuration. In such a case, the distribution function for the density of states is Gaussian, as was shown in the context of the semiclassical treatment of electron-transfer reactions in Section III.B:

$$D_{oc} = dN_D(E)/dE \, N_D = (4\lambda kT)^{-1/2} \exp[-(E - E_{oc})^2/4\lambda kT] \tag{133}$$

and

$$D_{unoc} = dN_A(E)/dE \, N_A = (4\lambda kT)^{-1/2} \exp[-(E - E_{unoc})^2/4\lambda kT] \tag{134}$$

Figure 16 gives a graphical illustration of the two density of state functions. (The concentration of D was assumed to be equal to that of D^+, as is the case under standard conditions.) The curves are bell shaped with maxima at E_{oc} and E_{unoc} for occupied and unoccupied levels, respectively.

C. Interfacial Electron-Transfer Kinetics, Empirical Rate Laws for Electrochemical Reactions, and Comparison with Theory

1. Conducting Electrodes

Consider a simple electrochemical reaction on a conducting electrode which results in the conversion of an oxidized species O to the reduced product R (Figures 17 and 18):

$$n_{e^-}(\text{electrode}) + O \rightleftharpoons R \tag{135}$$

The experimentally observed quantity is the Faradaic current i (A/cm^2) which is directly proportional to the molar flux of the electroactive species at the surface of the electrode:

$$i = -\text{flux} \times F \times n_{e^-} \tag{136}$$

where

$$\text{flux} = -\frac{dn_O}{dt\,S} = \frac{dn_R}{dt\,S} \tag{137}$$

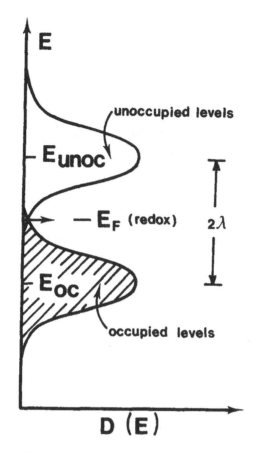

FIGURE 16. Gaussian distribution of occupied and unoccupied electronic states in a redox electrolyte. The intersection of the two curves corresponds to the Fermi level.

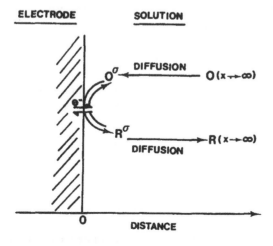

FIGURE 17. Schematic illustration of electron transfer between a metal and a solution redox couple (oxidized form O, reduced form R). The superscript σ indicates that the electroactive species is located at the electrode surface.

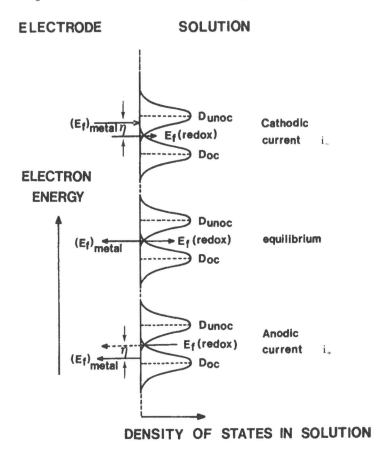

FIGURE 18.   Energetics for interfacial electron transfer from a metal to an acceptor A in solution. i_+ is the anodic and i_- is the cathodic current.

F is Faraday's constant and n_e^- is the number of electrons transferred during the reduction of the acceptor. S is the surface area of the electrode. The net current is composed of a cathodic (i_-) and an anodic (i_+) contribution:

$$i = i_+ + i_- \tag{138}$$

Assuming that the heterogeneous electron-transfer reaction is first order with respect to the concentration of electroactive species, the currents can be written as

$$i_+ = nFk_+ c_R^\sigma \tag{139}$$

and

$$i_- = nFk_- c_O^\sigma \tag{140}$$

where the superscript σ indicates concentrations in the close vicinity of the electrode surface. k_+ and k_- are the heterogeneous rate constants, expressed in centimeters per second, for the oxidation and reduction process, respectively. In the vast majority of electrochemical reactions these rate constants depend exponentially on the overpotential used to drive the electron-transfer process. Therefore, k_+ and k_- are expressed in terms of the Tafel equations:

$$k_+ = k_+^\circ \exp[(1 - \alpha)nFE/RT] \tag{141}$$

and

$$k_- = k_-^\circ \exp(-\alpha nFE/RT) \qquad (142)$$

where α is the transfer coefficient for the cathodic reaction, and E is the electrode potential measured against a reference electrode. Introducing the overvoltage

$$\eta = E - E_{eq} \qquad (143)$$

and expressing the equilibrium potential by the Nernst equation

$$E_{eq} = E^\circ + \frac{RT}{nF} \ln \frac{c_O^\sigma}{c_R^\sigma} \qquad (144)$$

gives the rate constants:

$$k_+ = k_+^\circ (c_O^\sigma/c_R^\sigma)^{1-\alpha} \exp[(1 - \alpha)nFE^\circ/RT] \times \exp[(1 - \alpha)nF\eta/RT] \qquad (145)$$

and

$$k_- = k_-^\circ (c_R^\sigma/c_O^\sigma)^{\alpha} \exp(-\alpha nFE^\circ/RT) \times \exp(-\alpha nF\eta/RT) \qquad (146)$$

which after insertion into Equations 134 and 140 yield the net current:

$$i = nF(c_O^\sigma)^{1-\alpha} (c_R^\sigma)^{\sigma} k_+^\circ \exp[(1 - \alpha)nFE^\circ/RT] \times \exp[(1 - \alpha)nF\eta/RT]$$
$$- k_-^\circ \exp(-\alpha nFE^\circ/RT) \times \exp(-\alpha nF\eta/RT) \qquad (147)$$

If the system is at equilibrium ($\eta = 0$),

$$i = i_+ + i_- = 0$$

which implies that

$$k_+^\circ \exp[(1 - \alpha)nFE^\circ/RT] = k_-^\circ \exp(-\alpha nFE^\circ/RT)] = k^\circ \qquad (148)$$

the exchange current i_O which flows at equilibrium is therefore

$$i_O = (i_+)_{eq} = -(i_-)_{eq} = nF(C_O)^{1-\alpha} (C_R)^{\alpha} \cdot k^\circ \qquad (149)$$

and Equation (147) can be written as

$$i = i^\circ(\exp[(1 - \alpha)nF/RT] - \exp[-\alpha nF/RT]) \qquad (150)$$

which is the Butler-Volmer relation. These considerations show that a heterogeneous electron-transfer process at a conducting electrode is fully characterized by two kinetic parameters: the transfer coefficient α and the rate parameter k_O.

It is instructive to compare these empirical equations to the predictions of electron-transfer theory. The effect which is of interest to us is the change in the rate constant of the electron transfer upon varying the driving force, i.e., the overvoltage of the reaction. For simplicity we shall consider the electron transfer only in one direction, say, from the electrode to the acceptor O. Furthermore, since in electron-transfer theory the driving force refers to standard-state conditions it is necessary to select the concentrations such that $c_O = c_R$. Under standard-state conditions the overvoltage becomes

$$\eta^\circ = E - E^\circ \tag{151}$$

The expression for the rate constant of the cathodic process reduces to

$$k_- = k_-^\circ \exp(-\alpha nF\eta^\circ/RT) \tag{152}$$

yielding for the derivative

$$d \ln k_-/d\eta^\circ = -\alpha nF/RT \tag{153}$$

If the same derivative is formed from the Marcus equation (Equation 46) one obtains

$$d \ln k_{et}/d\eta^\circ = nFd\ln k_{et}/d\Delta G^\circ = -\left(\frac{1}{2} + \frac{\Delta G^\circ}{2\lambda}\right)\frac{nF}{RT} \tag{154}$$

Thus, by comparison

$$\alpha = \frac{1}{2} + \frac{\Delta G^\circ}{2\lambda} \tag{155}$$

or, since $\Delta G^\circ = nF\eta^\circ - W_O + W_R$

$$\alpha = \frac{1}{2} + \frac{nF\eta^\circ}{2\lambda} + \frac{W_R - W_O}{2\lambda} \tag{156}$$

Here W_O and W_R are the work terms involved in the formation of the precursor and successor complex at the electrode surface, respectively. If these work terms are small compared to λ, or if they cancel each other, one obtains

$$\alpha = \frac{1}{2} + \frac{nF\eta^\circ}{2\lambda} \tag{157}$$

Experimentally, in a great number of electrochemical reactions on metal electrodes, Tafel behavior is observed.[39] Only exceptionally, have small variations of the transfer coefficient with overvoltage been reported.[40] It has been argued,[41] however, that in typical electro-chemical experiments the overvoltage is not varied over a wide enough range to provide a critical test for the validity of Equation 157.

2. Semiconducting Electrodes

A characteristic feature of the metal-solution interphase is that any change in the externally applied voltage will produce a corresponding change in the potential drop across the double layer:

$$dV = d\phi_{dl} \tag{158}$$

With semiconducting electrodes the situation is different inasmuch as a space charge layer is formed within the semiconductor when it is placed in contact with the solution. The nature of this space charge layer will be discussed in more detail in Chapter 3. Suffice it to say that the capacity of this layer is usually much smaller than that of the adjacent double layer in the electrolyte. Since the total capacity is given by

$$1/C = 1/C_{dl} + 1/C_{sc} \tag{159}$$

it follows that

$$1/C \approx 1/C_{sc} \tag{160}$$

For this reason, with semiconductor electrodes any variation in the externally applied voltage often changes only the potential drop within the semiconductor while the potential difference between semiconductor and solution is not affected.

In Figure 19 we consider the case of an n-type semiconductor in contact with an electrolyte containing the O-R redox couple. Figure 19A depicts the case where the system is at equilibrium. A depletion layer has been formed within the semiconductor by equilibration of the Fermi levels of the electrode and the solution. In Figure 19B a cathodic bias is applied to raise the Fermi level in the semiconductor. While this reduces the band bending it does not affect the potential drop across the double layer. The driving force for electron transfer is given by the difference between the conduction band edge position at the semiconductor surface, i.e., the flat band potential (V_{fb}) and the Nernst potential of the redox couple in solution:

$$\eta = V_{fb} - E_{eq} \tag{161}$$

Changing the bias voltage has no effect on η. Therefore, the rate constant for charge injection from the donor R in the conduction band of the semiconductor k_+ should be independent of the applied voltage. The same is true for the anodic current which is proportional to k_+ (Equation 139).

The situation is similar if the electron transfer occurs in the opposite sense, i.e., from the conduction band of the semiconductor to an acceptor in solution. Equation 161 remains applicable, and η does not vary with applied bias voltage if the capacity of the space charge layer is much smaller than that of the double layer in the electrolyte. However, for the cathodic process the fact needs to be taken into consideration that, due to the band bending, the concentration of electrons in the semiconductor bulk differs from that present at the surface where the electrochemical reaction occurs. This is accounted for by introducing the Boltzmann factor:

$$P = \exp[(\phi_{surface} - \phi_{bulk})F/RT] \tag{162}$$

into the expression for the cathodic curent (Equation 140). P expresses the ratio of the electron concentrations at the surface and in the bulk. Equation 140 becomes

$$i_- = -nFc_O^s \, k_- \, \exp[(\phi_{surface} - \phi_{bulk})F/RT] \tag{163}$$

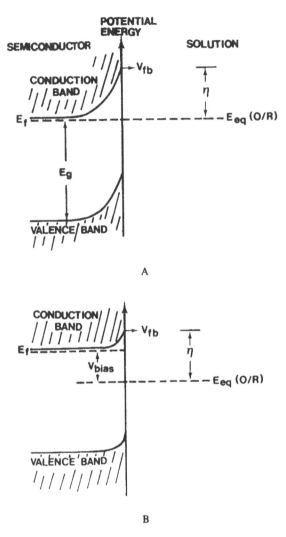

FIGURE 19. Potential distribution at the interface of an n-type semiconductor in contact with an electrolyte containing a redox couple (O/R). (A) System of equilibrium and (B) cathodic bias applied. The overvoltage for charge transfer across the interface is not affected by the change in bias potential.

where k_- is independent of the applied potential in contrast to metal electrodes. Thus, for electron transfer from the conduction band of an n-type semiconductor to an acceptor in solution, the transfer coefficient is 1 and the Tafel slope, i.e., the voltage increment necessary to increase the current by a factor of 10, is 59 mV at room temperature.

Analogous considerations apply to charge transfer reactions between p-type semiconductors and redox couples in solution. The cathodic current arising from hole injection in the valence band should be independent of the applied potential, while the anodic current, due to hole ejection in the electrolyte, should depend exponentially on the applied bias, the Tafel slope being 59 mV.

As for the photoactivated charge transfer reactions on semiconducting electrodes, these usually involve minority charge carriers produced by band gap excitation of the semiconductor. The overvoltage for interfacial transfer of minority carriers is

$$\eta = V_{fb} + E_g - E_{eq} \qquad (164)$$

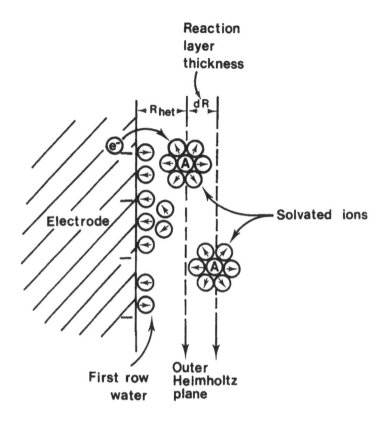

FIGURE 20. Schematic illustration of the individual steps involved in the interfacial electron transfer from an electrode to a solvated acceptor ion.

Since for $C_{sc} \ll C_{dl}$ η does not depend on the applied voltage, the heterogeneous rate constant for minority carrier transfer is also voltage independent. However, the photocurrent does depend on the bias potential. The latter affects the band bending within the semiconductor, and therefore also the efficiency of light-induced charge separation.

Finally, it should be noted that for semiconductors with a high density of surface states the condition $C_{sc} \ll C_{dl}$ is no longer valid. The reason for this behavior is that the surface-state capacity is parallel to that of the space charge layer, the total capacity of the semiconductor being the sum of both. If the surface states are sufficiently dense ($>10^{13}$ cm^{-2}), C_{sc} becomes of the same magnitude as C_{dl}, i.e., in the 10^{-4} F range. As a consequence, the flat band potential and the potential difference between the semiconductor and solution shift with applied bias voltage (Fermi level pinning). At a very high density of surface states the applied voltage drops only across the double layer and not within the space charge layer of the semiconductor. In such a case, metallike charge transfer kinetics will be observed.

D. Interpretation of the Preexponential Factor in Electrochemical Reactions

Experimental studies of the kinetics of heterogeneous charge transfer processes yield a rate constant with the units centimeters per second. On the other hand, the rate constant k_{et} calculated from the theory of electron-transfer processes is expressed in seconds. The value of the latter corresponds to the reciprocal average time required for interfacial electron transfer. In the following we establish a relation between these two rate parameters. As an example we consider the elementary steps involved in an electrochemical reaction, i.e., the reduction of an acceptor ion A (Figure 20).

First, the acceptor must diffuse from the solution bulk to the electrode surface. Since A is usually solvated and the electrode covered by a monolayer of chemisorbed solvent molecules, the approach to the surface is restricted to the distance of the outer Helmholtz plane. For the equilibrium between A present at the surface and in the bulk, we write

$$A_{bulk} \overset{k_d}{\underset{k_d^-}{\rightleftharpoons}} A_{surface} \tag{165}$$

and define the equilibrium constant as

$$K = \Gamma_A/c_A = k_d/k_d^- \tag{166}$$

where Γ_A (moles per square centimeter) is the surface concentration of the acceptor. Once A has reached the surface of the electrode, interfacial electron transfer can occur:

$$(A)_{surf} + e_{metal}^- \overset{k_{et}}{\longrightarrow} (A^-)_{surf} \tag{167}$$

The location of A during the redox event is within the reaction layer of thickness dR_{het}. The latter is of the same magnitude as the tunneling length β^{-1}, i.e., between 0.5 and 2Å.

Treating the sequence of Reactions 165 and 167 with the steady-state approximation for the surface concentration of A gives the desired relation between the observed rate constant for the cathodic process and that for interfacial electron transfer, i.e., k_- and k_{et}:

$$1/k_- = 1/k_d + 1/Kk_{et} \tag{168}$$

The rate constant for a diffusional encounter between a spherical reactant and a planar surface has been given as:[42]

$$k_d = 3D_A/2\Delta \tag{169}$$

where Δ is the average distance over which the acceptor is required to move between adjacent lattice positions in the solvent. Inserting typical values, i.e., $D_A = 10^{-5}$ cm²/sec and $\Delta = 3 \times 10^{-8}$ cm, in Equation 169 yields $k_d = 5 \times 10^2$ cm/sec. Frequently the product $K \cdot k_{et}$ in Equation 168 is much smaller than k_d, implying that the adsorption equilibrium is always established during the electron-transfer reaction. Under these conditions Equation 168 reduces to

$$k_- = K \cdot k_{et} \tag{170}$$

Also, since in this case there is no local concentration gradient of A in the vicinity of the electrode, $\Gamma_A = c_A dR_{het}$ and $K = dR_{het}$.

It is interesting to note that in the literature the observed rate constant for heterogeneous electron transfer is often expressed as the product of a preexponential factor A and an activation energy term:

$$k_- = A \exp(-\Delta G^*/RT) \tag{171}$$

A is assumed to be equal to k_d, i.e., the diffusion-limited rate constant for encounter of A with the electrode. The above considerations show this assumption to be incorrect. Thus, for very rapid heterogeneous electron-transfer reactions the steady-state approximation, Equation 168, must be employed. Only if the preequilibrium (Equation 165) is established during the redox events is Equation 171 applicable. However, the preexponential factor in such a case corresponds to the product $A = K \cdot v_O \cdot \kappa_{el}$ and not to k_d. The value of A depends on the electronic transmission factor κ_{el}. For an adiabatic reaction $\kappa = 1$. Assuming $v_o \approx 10^{13}$ sec^{-1}, one obtains $A = 10^5$ sec^{-1} which is several hundred times larger than the value of k_d.

E. Heterogeneous vs. Homogeneous Reorganization Energies

In this final section we will explore how the reorganization energy for the heterogeneous electron-transfer process, λ_{het}, is related to the reorganization energy observed in homogeneous reactions. As an example, we consider again the reduction of an electron acceptor A at an electrode and compare this to the homogeneous self-exchange reaction:

$$*A + A^- \rightarrow A + *A^- \tag{172}$$

where the star indicates that the molecule is isotopically marked. Intuitively, one would expect λ_{het} to be smaller than λ_{hom}. In fact, $\lambda_{het}/\lambda_{hom}$ should be approximately 0.5, since in the exchange reaction two reactants and the solvation shells are undergoing a rearrangement in forming the activated complex, while in the heterogeneous system there is but one such particle. This idea is indeed confirmed by the finding that the inner sphere component of λ_{het} follows in general the relation $\lambda_{het} = \frac{1}{2}\lambda_{hom}$. However, the relation between the outer sphere components of λ is more complicated since it is influenced by the distance R between the reacting molecules A and A$^-$ in the activated complex in the exchange reaction, and the distance R_{het} between A and the electrode in the heterogeneous process:

$$\lambda_{out}^{het} = \frac{1}{2}\lambda_{out}^{hom} + \frac{e^2}{2}\left(\frac{1}{R} - \frac{1}{2R_{het}}\right)\left(\frac{1}{n^2} - \frac{1}{D_s}\right) \tag{173}$$

Since $2R_{het}$ is either equal to or greater than R, depending on whether the reactant can or cannot penetrate the layer of solvent molecules present at the electrode surface, one finds $\lambda_{out}^{het} \geq \frac{1}{2}\lambda_{out}^{hom}$, and therefore

$$\lambda_{het} \geq \frac{1}{2}\lambda_{hom} \tag{174}$$

Equation 174 allows for a direct comparison of the rate constant for a homogeneous exchange reaction with that of an interfacial electron transfer involving the same redox couple. In the latter case it is convenient to refer to the rate constants at zero overvoltage (k_O) which are readily derived from measurements of the electrochemical exchange current densities. If the equality sign in Equation 174 applies, a plot of ln (k_o) against ln k_{ex} should give a straight line with a slope of 0.5. This has indeed been observed for a series of different redox couples.[43]

REFERENCES

1. **Ulstrup, J.**, *Charge Transfer Processes in Condensed Media*, Springer-Verlag, West Berlin, 1979; **Cannon, R. D.**, *Electron Transfer Reactions*, Butterworths, London, 1980.
2. **Atkins, P. W.**, *Molecular Quantum Mechanics*, Oxford University Press, Oxford, 1983.
3. **Gamov, G. Z.**, *Physik*, 51, 205, 1928.
4. **Zamaraev, K. I. and Khairutinov, R. F.**, *Chem. Phys.*, 4, 181, 1974.
5. **Marcus, R. A.**, *Annu. Rev. Phys. Chem.* 15, 155, 1964.

6. **Sutin, N.**, *Acc. Chem. Res.*, 15, 275, 1982; **Marus, R. A. and Sutin, N.**, *Biochim. Biophys. Acta*, 811, 265, 1985.
7. **Sutin, N.**, in *Tunneling in Biological Systems*, Chance, B., de Vault, D. C., Frauenfelder, H., Marcus, R. A., Schrieffer, J. R., and Sutin, N., Eds., Academic Press, New York, 1979, 201,
8. **Hopefield, J. J.**, *Proc. Natl. Acad. Sci. U.S.A.*, 71, 3640, 1974.
9. **Alexandrov, I. V., Kaimtimov, K. I., and Zamaraev, O.**, *Chem. Phys.*, 32, 123, 1978.
10. **Beitz, J. V. and Miller, J. R.**, *J. Chem. Phys.*, 71, 4579, 1979.
11. **Hupp, J. T. and Weaver, M. D.**, *J. Phys. Chem.*, 88, 1463, 1984; *J. Electroanal. Chem.*, 152, 1, 1983.
12. **Li, P. and Weaver, M. J.**, *J. Am. Chem. Soc.*, 106, 6107, 1984.
13. **Sugi, M., Nembach, K., and Möbius, D.**, *Thin Solid Films*, 27, 205, 1975.
14. **Nielson, R. M. and Wherland, S.**, *J. Am. Chem. Soc.*, 107, 1505, 1985.
15. **Miller, J. R.**, *Nuovo J. Chim.*, 11, 82, 1987.
16. **McConnell, H. M.**, *J. Chem. Phys.*, 35, 508, 1961.
17. **Landau, L.**, *Phys. Z.Sowjetunion*, 1, 88,1932.
18. **Förster, T.**, *Naturwissenschaften*, 33, 166, 1946.
19. **Dexter, D. C.**, *J. Chem. Phys.*, 21, 836, 1953.
20. **Hopefield, J. J.**, *Biophys. J.*, 18, 311, 1977.
21. **De Vault, D.**, *Quantum-Mechanical Tunnelling in Biological Systems*, Cambridge University Press, New York, 1984.
22. **De Vault, D.**, *Rev. Biophys.* 13, 387, 1980.
23. **Huang, K. and Rhys, A.**, *Proc. R. Soc. London Ser. A*, 204, 406, 1950.
24. **Dogonadze R. R., Ulstrup, J., and Kharkats, Yu.**, *J. Theor. Biol.*, 40, 279, 1973.
25. **Levich, V. G. and Dogonadze, R. R.** *Dokl. Akad. Nauk SSSR*, 123, 123, 1959; **Levich, V. G. and Dogonadze, R. R.**, *Coll. Czech. Chem. Commun.*, 26, 193, 1961.
26. **Jortner, J.**, *J. Chem. Phys.*, 64, 4860, 1976.
27. **Kubo, R.**, *Phys. Rev.*, 86, 429, 1952.
28. **Lax, M.**, *J. Chem. Phys.*, 20, 1752, 1952.
29. **Bockris, J. O. M. and Khan, S. U. M.**, *Appl. Phys. Lett.* 42, 124, 1983.
30. **Gerischer, H.**, in *Physical Chemistry: An Advanced Treatise*, Vol. 9A, Henderson, D. and Jost, W., Eds., Academic Press, New York, 1970.
31. **Meming, R.**, in *Electronanalytical Chemistry*, Vol. 11, Bard, A. J., Ed., Marcel Dekker, New York, 1979.
32. **Khan, S.U.M. and Bockris, J.O.M.**, *J. Phys. Chem.*, 87, 2599, 1983.
33. **Lohman. F.**, *Z. Naturforsch. Teil A*, 22, 813, 1956.
34. **Gomer, R. and Tryson, G.**, *J. Chem. Phys.*, 66, 4413, 1977.
35. **Trasatti, S.**, *J. Electroanal. Chem.*, 139, 1, 1982.
36. **Trasatti, S.**, *J. Electroanal. Chem.*, 209, 417, 1986.
37. **Kittel, C.**, *Introduction to Solid State Physics*, John Wiley & Sons, New York, 1976.
38. **Khan, S.U.M. and Bockris, J.O.M.**, *J. Phys. Chem.*, 88, 2504, 1984.
39. **Conway, B. E., Bockris, J.O.M., Yeager, E., Khan, S.U.M., and White, R.**, Eds., *Comprehensive Treatise of Electrochemistry*, Vol. 7, Plenum Press, New York, 1983.
40. **Saveant, J. M. and Tessier, D.**, *Discuss. Faraday Soc.*, 74, 1982.
41. **Marcus, R. A.**, *J. Chem. Phys.*, 67, 853, 1963.
42. **Reiss, H. J.**, *J. Chem. Phys.*, 18, 996, 1950.
43. **Frese, K.W.**, *J. Phys. Chem.*, 85, 3911, 1981.

Chapter 2

LIGHT ENERGY HARVESTING AND PHOTOINDUCED ELECTRON TRANSFER IN ORGANIZED MOLECULAR ASSEMBLIES

I. INTRODUCTION

The world faced an acute energy crisis 15 years ago. A sharp increase in the price of oil and dwindling supplies of fossil resources rendered the scientific community aware of the necessity of exploring possibilities for new energy supplies. The use of sunlight to produce alternative and environmentally clean fuels, such as hydrogen, appeared as an attractive strategy, and a large interdisciplinary effort was soon begun.

Scientists knew, of course, that in striving to achieve this goal much could be learned from photosynthesis. Nature has built up a fascinating device to make use of sunlight in order to drive a thermodynamically uphill reaction, i.e., the reduction of carbon dioxide to carbohydrates by water. The photosynthetic units assembled in the thylakoid membranes comprise antenna pigments for light energy harvesting and a reaction center consisting of two photosystems in series. Several key features as to how photosynthetic energy conversion operates are known. Light-induced charge separation is achieved through judicious spatial arrangements of the chlorophyll (Chl) and elements of the electron-transport chain across the membrane. Cooperative interaction between these components allows the electron transfer to proceed in a vectorial fashion: positive charges are accumulated at the inside of the vesicles, while the negative countercharges are transferred to the outer surface. Enzymes play the role of catalysts that couple the charge separation events to fuel-generating reactions, i.e., the oxidation of water to oxygen and reduction of NAD^+ to NADH.

While artifical photoconversion devices should not attempt to imitate all the intricacies of natural photosynthesis, it is inconceivable that the challenging task of driving endergonic chemical reactions such as the cleavage of water into hydrogen and oxygen could be accomplished without suitable engineering on the molecular level. Microheterogeneous solution systems have therefore become a topic of primary interest. The simplest equivalent of a biological membrane is a surfactant micelle. These are aggregates of approximately spherical structure which form spontaneously in aqueous solutions of ionic or nonionic detergents. The hydrocarbon tails of the latter form the interior, and the polar head groups form the surface of these assemblies. Typically, about 60 to 100 detergent molecules are associated in one micelle. In the case of ionic micelles there exists an electrostatic potential gradient extending from the surface into the surrounding aqueous solution. Thus, some of the most important features of biological membranes, such as the presence of microscopically small hydrophobic regions as well as the charged lipid water interface, are mimicked by micellar assemblies. We shall illustrate in the following sections how these simple model aggregates can serve as a powerful tool for controlling the kinetics of energy and electron-transfer events.

II. TRANSFER OF ELECTRONIC EXCITATION ENERGY IN MICELLAR ASSEMBLIES

An important feature of micellar assemblies is the microheterogeneous character; the micellar interior presents a hydrophobic region of minute dimensions which is separated from the aqueous bulk phase by the palisade layer constituted by the polar head groups of the surfactants. As a consequence, energy and electron-transfer reactions involving solubilized substrates cannot be treated by the conventional kinetic laws that are applied in

EXCITATION ENERGY TRANSFER IN MICELLAR SYSTEMS

FIGURE 1. Schematic illustration of intramicellar energy transfer from an optically excited donor (D*) to an acceptor.

homogeneous solution kinetics. These events occur within a reaction space of very small dimension and involve only a few sequestered molecules. Therefore, a stochastic approach is necessary in the analysis of the kinetic events. In addition, since the reactions to be considered occur, in general, very rapidly there is no time for exchange of molecules between different micelles. In such a situation, the statistics of reactant distribution over the host aggregates play an important role. During the last decade, new kinetic concepts were developed which took into account these unique features and provided for adequate models to interpret the experimental observations. The most important cases relating to intramicellar energy transfer will now be discussed.

A. Intramicellar Triplet Energy Transfer

In this section we consider the kinetics of a process in which a species confined to a micellar assembly is photoexcited to a triplet state and subsequently transfers electronic excitation energy to a partner molecule that is present in the same micelle (Figure 1):

$$D^T + A^S \rightarrow D^S + A^T \tag{1}$$

Such a reaction proceeds via an electron exchange mechanism which requires a strong overlap of the donor and acceptor wave functions and hence close approach of the reactants. An entirely analogous situation prevails in the case of intramicellar electron transfer. We recall from Chapter 1 that the efficient coupling of the wave functions of electron donor and acceptor is also a prerequisite for the occurrence of rapid charge transfer. Hence, we expect — and this has meanwhile been experimentally confirmed — that the rate laws for both intramicellar electron and triplet energy transfer are similar.

In the first system examined, the triplet energy of the acceptor is considerably smaller than that of the donor and reaction 1 occurs irreversibly. Apart from the simpler kinetic analysis, such a case is particularly suited to illustrate the statistical laws that govern the probe distribution over the micelles. Both energy donor and acceptor are hydrophobic species and hence will be associated almost exclusively with the micellar aggregates. Prior to the photoexcitation the micelle-solubilizate system is at equilibrium, i.e., the concentration of micelles remains constant, and the distribution of the probes among the micelles is stationary. Interactions between D and A are assumed to be negligible, and thus their respective distributions among the micelles may be considered independent. We consider the case where the donor molecule is excited by an ultrashort laser pulse and are interested in the time evolution of the system following light excitation. Furthermore, we note that the fraction of micelles containing more than one triplet is negligible. Therefore triplet-triplet annihilation does not need to be considered.

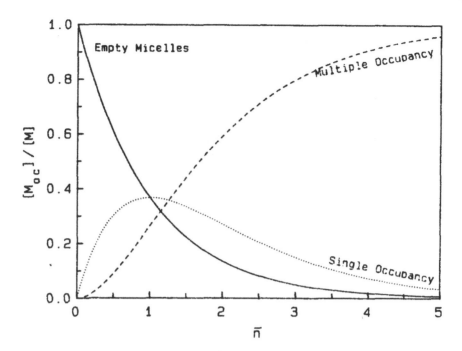

FIGURE 2.   Graphic illustration of the Poisson distribution for calculation of the number of solubilizate molecules associated with a micelle as a function of the average occupancy.

Since the micelles contain at most a few acceptor molecules, these are distributed in a statistical fashion over the host aggregates. We use the Poisson equation to express this distribution. According to this law the fraction of micelles having i probe molecules incorporated is given by

$$\frac{M_i}{M} = \frac{\bar{n}^i}{i!} \exp(-\bar{n}) \tag{2}$$

where $\bar{n}$ is the average occupancy of the micelles. The latter can be calculated from the concentrations of acceptor A and surfactant via

$$\bar{n} = \frac{[A] \cdot v}{[surfactant] - CMC} \tag{3}$$

where v is the aggregation number and CMC the critical micelle concentration. In Figure 2 we graphically illustrate the Poisson distribution law. It is predicted, for example, that for an average occupancy of one, only one third of the aggregates actually contain one acceptor. The remainder is divided up evenly into empty micelles and those containing more than one solubilizate.

When applying Equation 2, the assumption is made implicitly that there is an unrestricted number of solubilization sites available in a micelle. In most experimental situations the micellar occupancies by probe molecules are small, and the Poisson distribution is closely obeyed. However, deviations are expected to occur in the case where a large number of probe molecules, approaching the solubilization limit, is present in the micelles.

Returning now to the analysis of intramicellar triplet energy transfer, we realize that in solution there are micelles with different D and A association. For a particular type of micelle containing j donor and i acceptor molecules Equation 1 may be reformulated as

$$D^T + (j - 1)D + iA \xrightarrow{i \times k_q} A^T + jD + (i - 1)A \qquad (4)$$

In the specific case where $i = j = 1$, a micelle contains only one D^T and A molecule each after excitation, and subsequent energy transfer involves a reactant pair. As will be shown below, such a process follows first-order kinetics. The rate constant k_q has the unit seconds and expresses the reciprocal mean time required for the transfer of excitation energy from the donor to the acceptor molecule. To evaluate experimental data micelles with multiple A and/or D associations must be considered. The specific rate of energy transfer in the forward direction is expected to increase linearly with the occupation number of A.[1] Thus, the differential time law for triplet states in micelles with A associations is

$$-\frac{d[D_i^T]}{dt} = i \times k_q[D_i^T] + k_d[D_i^T] \qquad (5)$$

where k_d is the rate constant for triplet deactivation in micelles that do not contain any acceptor. The solution of Equation 5 is a simple exponential:

$$[D_i^T] = [D_i^T](0) \exp[-(k_d + i \times k_q)t] \qquad (6)$$

where $[D_i^T]$ (0) is the concentration of triplet states immediately after light excitation that occupy micelles with i acceptor associations. From Equation 2, one obtains

$$[D_i^T](0) = [D^T](0) \frac{\bar{n}_A^i \exp(-\bar{n}_A)}{i!} \qquad (7)$$

The time law for the total triplet concentration,

$$[D^T] = \sum_{i=0}^{i=\infty} [D^T] \qquad (8)$$

derived from Equations 6 to 8 is

$$[D^T] = [D^T](0) \exp(-k_d t + n_A[\exp(-k_q t) - 1]) \qquad (9)$$

In most experimental systems $k_d \ll k_q$ due to the protective effect of the micellar environment with regard to nonradiative triplet deactivation.[2] In this case, Equation 9 predicts a two-step behavior for the triplet-state decay: a fast component due to intramicellar energy transfer to the acceptor followed by a slower first-order process arising from micelles that contain only donor molecules. If the assumption of a Poisson distribution of acceptor molecules over the micellar aggregates is correct, then the ratio of triplet-state concentration present after completion of the fast decay and that present initially should be given by

$$\frac{[D^T](\text{plateau})}{[D^T](0)} = \exp(-\bar{n}_A) \qquad (10)$$

These predictions were tested by using *N*-methylphenothiazine, as an energy donor and *trans*-stilbene as an acceptor;[3] these were solubilized in aqueous solutions of cetyltrimethyl-

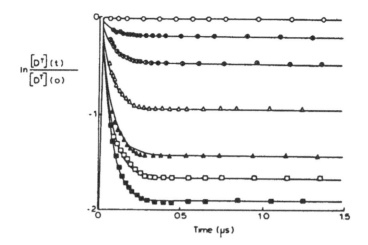

FIGURE 3. Kinetics of the irreversible energy transfer in micellar CTAB (2.10^{-2} M) solution from N-methyl-phenothiazine (MTPH) triplet to *trans*-stilbene. MPTH = $1.6.10^{-4}$. (*trans*-stilbene: ○, 0.0 M; ●, 4.10^{-5} M; ◑, 1.10^{-4} M; △, 2.10^{-4} M; ▲ 3.10^{-4} M; □, $3.5.10^{-4}$ M; and ■, 4.10^{-4} M). The curves are calculated according to Equation 9.

ammonium chloride (CTAC) micelles. The gap between the triplet levels of N-methylphenothiazine, MPTH, (E_T = 256.2 kJ/mol) and that of *trans*-stilbene (E_T = 209 kJ/mol) is sufficiently large to ensure that Reaction 1 occurs irreversibly. MPTH was excited by a frequency-doubled ruby laser, and the time course of the MPTH-triplet absorption decay was monitored by kinetic spectroscopy. Concentration-time profiles for $MPTH^T$ in the short time domain corrected for the slower decay component are shown in Figure 3. It is noted that an increase in the *trans*-stilbene concentration induces a decrease in $MPTH^T$ concentration in the plateau region. The detailed analysis showed that Equation 10 is obeyed, which provides experimental support for the validity of Poisson statistics for the distribution of stilbene molecules over the CTAC micelles. The micellar aggregation number derived from these studies is 92, in excellent agreement with literature results.

The specific rate for intramicellar triplet energy transfer is obtained from kinetic evaluation of the decay curves. The solid lines in Figure 3 were calculated from Equation 9, and an optimal fit with the experimental data was obtained for k_q = $(1.5 \pm 0.2) \cdot 10^7$ sec^{-1}. This value implies that the average time for energy transfer from the MPTH donor to the *trans*-stilbene acceptor species solubilized within a CTAB micelle is 67 nsec. In view of the excellent agreement between the experimental data and the predictions of Equation 9, we can accept with confidence the premises that lead to the derivation: (1) the description of the energy-transfer process by a set of first-order rate laws where the rate constant increases linearly with the number of acceptor molecules present in the host aggregates and (2) the Poisson distribution of the acceptor among the micelles.

It is interesting to note that in homogeneous solution, e.g., an ethanol-water mixture (1:1, v/v), the same energy transfer from triplet-excited MPTH to *trans*-stilbene occurs with a second-order rate constant of k_q = $(3.7 \pm 0.3) \cdot 10^9$ mol^{-1} sec^{-1}. As this value corresponds to the limit expected for diffusion-controlled reactions it can be assumed that the energy transfer in the micellar system proceeds also on every encounter of the reactants. Hence, the mean energy-transfer time of 67 nsec may be interpreted as the average time between two successive collisions of a donor-acceptor pair in the micelle.

In order to rationalize these experimental observations, we developed a general model[1] which explored the role of dimensionality and spatial extent in intramicellar energy- and electron-transfer reactions. Using the continuum theory of reaction diffusion kinetics, it was

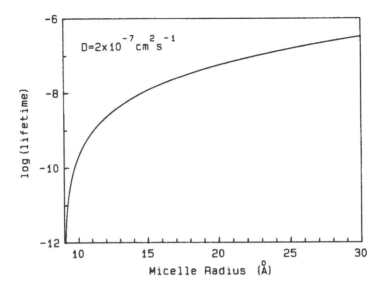

FIGURE 4.   Graphic illustration of the effect of micellar radius on the mean encounter time between a pair of reactants, $D = 2 \times 10^{-7}$ cm²/sec, $R\mu = 4$ Å.

possible to show that the reaction is expected to follow to a very good approximation a first-order rate law when only one donor and one acceptor species associated with the micelle. The linear increase in the first-order rate constant with the occupancy of the micelle also emerged from this model. Furthermore, from this treatment it was possible to derive an approximate expression for the average time required for diffusional encounter of a pair of reactants in a micelle, assuming that one of the reactants is fixed in the micellar center:

$$\tau = \frac{r_0^3 \left(1 - \dfrac{a}{r_0}\right)^2}{3Da} \tag{11}$$

where D is the mean diffusion coefficient of the two reactant molecules in the micelle, a the sum of the reaction radii, and $r_0 = R_m - R_s$ the difference between the micellar radius and that of one reactant, i.e., the donor. Figure 4 illustrates the dependency of the collisional time between a pair of reactants on the micellar radius for $D = 2 \times 10^{-7}$ cm²/sec. Using $r_0 = 15$ Å, $a/r_0 = 0.2$, one obtains $\tau = 150$ nsec, which is in fair agreement with the experimental results.

A particularly interesting situation which was also examined recently[3] is the case where the triplet levels of the donor and acceptor are so close that Reaction 1 occurs reversibly. For example, if in the case of MPTH as the donor naphthalene is employed as an acceptor, the energy difference is only -0.4 kJ/mol, and after excitation of MPTH the system relaxes to the equilibrium state:

$$\text{MPTH}^T + \text{naphthalene} \underset{k_{-q}}{\overset{k_q}{\rightleftharpoons}} \text{MPTH} + \text{naphthalene}^T \tag{12}$$

The mathematical treatment for intramicellar reversible energy transfer is similar to the irreversible case[3] and hence will not be presented here. From laser photolysis investigations of the MPTH-naphthalene system in cetyltrimethylammonium bromide (CTAB) micelles, the values of k_q and k_{-q} obtained were $(2.8 \pm 0.5) \cdot 10^6$ and 3.3×10^6 sec^{-1} respectively.

Clearly, in this case the rate of energy transfer is smaller than for the MPTH^T-stilbene reaction. Such behavior is expected, since the rate constants for isoenergetic energy-transfer processes are usually below the limit for diffusion-controlled reactions.[4] This implies that not every encounter between a donor and acceptor molecule within the micelle will be reactive.

In the comparison of triplet energy-transfer processes in micellar assemblies with the same reaction in homogeneous solution, two important features emerge: (1) through confinement of the reaction space to micellar size, both a drastic enhancement of the observed transfer rate and a reduction of the reaction order from 2 to 1 is obtained; and (2) apart from the equilibrium constant, statistical factors, i.e., the distribution of donor and acceptor over the host aggregates, determine to what extent the reaction will occur.

In closing this section, it should be pointed out that Equation 9 can also be directly obtained from the general expression derived for the kinetics of excited-state quenching processes in micellar assemblies[4] where allowance was made for the quencher to move freely between the micellar and aqueous phases. The kinetics with which we are concerned in the case of triplet energy transfer are simplified by the fact that the location of the quencher is restricted to the micellar phase.

Apart from intramicellar triplet energy transfer, Equation 9 has been widely used to describe luminescence quenching in micellar solutions in general. It is assumed that both fluorophor and quencher are quantitatively associated with the micelles. Yekta et al.[5] have suggested that such experiments offer a convenient way of determining micellar aggregation numbers via steady-state luminescence analysis. To deal with these interesting quenching experiments in more detail would be beyond the scope of this monograph. For further discussion, the interested reader is referred to pertinent published work.[6-14]

B. Triplet-Triplet Annihilation in Confined Reaction Space

The interaction of two triplet sensitizer molecules in solution gives rise to triplet-triplet annihilation. This process may lead to the formation of an excited singlet molecule ^{1}S* and ultimately to P-type delayed monomer fluorescence, or may result in the formation of a singlet excimer ^{1}D* responsible for the delayed excimer fluorescence. The formation of a triplet and the radiationless deactivation of the triplet pair can also be envisaged. The different pathways for triplet-triplet annihilation may be represented by

$$^3S + {}^3S \rightarrow ({}^3S \cdot {}^3S) \rightarrow {}^1S + {}^1S^* \quad \text{or} \quad {}^1D^* \tag{13}$$
$$\downarrow \qquad \searrow$$
$$^1S + {}^3S \qquad {}^1S + {}^1S$$

With respect to intramicellar triplet energy transfer discussed in Section II.A, we note that the intramicellar triplet-triplet annihilation process requires a different kinetic approach. In the former case, one triplet donor reacted with an ensemble of ground-state acceptors, whereas, in the latter case, the interactions in an ensemble of molecules in the triplet state must be considered. (Figure 5).

The partitioning of the triplets over small volumes and the fact that there is only a small number of these species present in each volume requires the application of a stochastic model for chemical reactions. Several authors have dealt with this problem.[15-23] In particular, McQuarrie et al.[15-17] have solved the stochastic model for the second-order reaction $2A \rightarrow C$, the reactants being compartmentalized in a single volume. A similar model is adequate to describe the present situation. Let us define $p_x(t)$ as the probability that a micelle contains x triplet molecules at time t. We have

$$p_x(t) = M_x(t)/M \tag{14}$$

FIGURE 5.   Triplet-triplet annihilation in micellar reaction space. Interactions between all excited species must be taken into account in order to calculate the reaction probability.

where $M_x(t)$ is the concentration at time t of the micelles with x triplet hosts. M is the total micelle concentration. We suppose that the triplets react in a pairwise fashion according to the reaction

$$^3S + {}^3S \xrightarrow{\text{k}} \text{products} \tag{15}$$

and that no triplet formation occurs as a result of the annihilation reaction. The parameter k_a represents the first-order rate constant for the annihilation reaction in micelles containing two triplets. In micelles containing x triplets, the rate for the transition $x \to (x - 2)$, i.e., the triplet disappearance rate, is assumed to be given by $\frac{1}{2}kx(x - 1)$. Among the x triplets, one can react with the $x - 1$ others, and $\frac{1}{2}x(x - 1)$ is the number of ways that a pair of reactant molecules can be chosen from a total of x reactant molecules. These hypotheses lead to the following set of differential difference equations:

$$dp_x(t)/dt = \tfrac{1}{2}k(x + 2)(x + 1)\,p_{x+2}(t) - \tfrac{1}{2}kx(x - 1)\,p_x(t) \tag{16}$$

where $x = 0, 1, 2,...$ and k is the rate constant. These equations are subject to the boundary conditions

$$p_x(0) = [(\bar{n}a)^x/x!]e^{-\bar{n}a} \quad \text{at time} \quad t = 0 \tag{17}$$

which arise from the statistics of probe distribution over the micelles. The parameter a = $[^3S(t = 0)]/[S(t = 0)]$ represents the fraction of sensitizer molecules excited to the triplet state. From this, it is possible to calculate an expression for the time evolution of the triplet concentration in the irradiated solution:

$$\frac{[TP(t)]}{[TP(0)]} = \frac{1}{\bar{n}a} \sum_{n=1}^{\infty} B_n \exp\{-\tfrac{1}{2}kn(n - 1)t\} \tag{18}$$

where $n = 1, 2, 3,...$ and

$$B_n = \frac{2n - 1}{2^n} e^{-\bar{n}a} \sum_{j=n}^{\infty} \frac{(\bar{n}a)^j}{(j - n)!} \cdot \frac{\Gamma\left(\dfrac{j - n + 1}{2}\right)}{\Gamma\left(\dfrac{j + n + 1}{2}\right)} \tag{19}$$

with $j = n, n + 2, n + 4,...$ Γ is the gamma function. For $t \to \infty$, Equation 18 reduces to

$$\frac{[TP(t \to \infty)]}{[TP(t = 0)]} = \frac{1 - e^{-\bar{n}a}}{\bar{n}a} \tag{20}$$

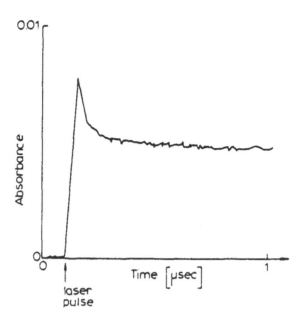

FIGURE 6. Kinetics of intramicellar triplet-triplet annihilation studied by laser photolysis of 1-bromonaphthalene in CTAB micellar solution (average occupancy n = 1.45). There is transient absorbance decay at 425 nm where the bromonaphthalene triplet shows an absorption.

The predictions of this model have been tested with 1-bromonaphthalene in micellar solutions of CTAB.[24] Intramicellar triplet-triplet annihilation is observed when solutions with high average micellar occupancy of bromonaphthalene are exposed to a laser pulse. Similar observations have been made by Lachich et al.[25] with $Ru(bipy)_3^{2+}$ solutions in sodium dodecyl sulfate (SDS) micelles. An oscillogram showing the temporal behavior of the bromonaphthalene triplet absorption at 425 nm is displayed in Figure 6. Here, a fast and partial decay of the triplet is followed by a plateau region which corresponds to a second, much slower decrease. The fast component of the decay is attributed to intramicellar triplet-triplet annihilation. Quantitative analysis showed that the predictions of the rate law in Equation 18 are in excellent agreement with the experimental results. This evaluation yields for the rate constant of intramicellar triplet-triplet annihilation a value of $k = (2.8 \pm 0.3) \cdot 10^7$ sec^{-1} The value of $k/2$ (two triplets annihilated per encounter) is of the same order of magnitude as the rate constants for irreversible intramicellar energy transfer discussed in the preceding section.

A plausible explanation for this similarity would be that, for both processes, an average diffusion time of 60 to 100 nsec, depending on the size of the reactant, must elapse before an encounter complex of the two reactant molecules is formed. The subsequent annihilation or energy-transfer reaction takes place with unit efficiency. Hence, this type of investigation provides another method of obtaining information on the diffusional displacement of probe molecules in molecular assemblies.

The fact that the bromonaphthalene triplet in Figure 3 does not decay completely via intramicellar triplet-triplet annihilation, but decays only partially to a plateau value, is a consequence of the statistical nature of the triplet distribution over the surfactant assemblies. Equation 17 describes the initial micellar occupancy by probe triplets. In the case of bromonaphthalene, intramicellar triplet-triplet annihilation does not result in any appreciable triplet reformation. Therefore, in micelles which initially contained an even number of triplets, triplet-triplet annihilation will be complete. On the other hand, if a micelle initially

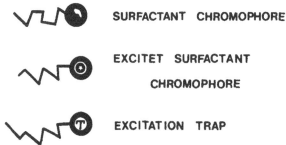

SURFACTANT CHROMOPHORE

EXCITET SURFACTANT
CHROMOPHORE

EXCITATION TRAP

FIGURE 7. Electronic excitation transfer within a micellar array of chromophores.

contained an uneven number of triplets, triplet-triplet annihilation will be incomplete and one triplet state per micelle will be left behind. The latter undergoes a first-order deactivation which is much slower than intramicellar triplet-triplet annihilation and hence barely visible on the time scale of observation used in Figure 6. Equation 20 gives a quantitative prediction of the concentration of triplet present after completion of intramicellar triplet-triplet annihilation. Experimental results are in excellent agreement with these predictions, vindicating the application of this statistical model to intramicellar triplet-triplet annihilation.

C. Excitation Energy Conduction Along a Micellar Surface to a Reactive Trap

We shall now discuss the case where excitation energy is conducted along the micellar surface to a reactive trap. Such a situation is particularly intriguing, since this process mimics the light energy harvesting in photosynthesis. Recall that plants absorb sunlight via the antenna pigments consisting of Chl and carotenoids. The excitation energy is subsequently transferred via a hopping process along the thylakoid vesicle surface to the reaction centers, where a light-induced charge separation takes place. There are about 500 antenna molecules for each reaction center. A micellar assembly imitating the biological unit can be conceived in the following way. A functional surfactant is used to form the micelles whose head group is constituted by the sensitizer S. Incorporated in the micellar surface is a host molecule (T) which functions as an energy trap (Figure 7). The study of the quenching of the S* luminescence by T should yield information on the energy-transfer rate within the micellar surface.

We model the kinetics of such an energy-transfer reaction by considering a sequence of elementary steps:

$$S^* \xrightarrow{P_{tr}} S \xrightarrow{P_{tr}} S \xrightarrow{P_{tr}} \dots B \qquad (21)$$
$$\downarrow P_d \qquad \downarrow P_d \qquad \downarrow P_d$$

which involve hopping of excitation energy from a chromophore to an adjacent one. The excitation is quenched when it reaches a B site, i.e., a chromophore located next to the quencher. If a quencher has n neighbors the number of A molecules which are available for energy conduction is

$$A = A_O - nQ \qquad (22)$$

where A_O and Q are the average number of A_O and quencher molecules per micelle, respectively.

In Equation 21 P_{tr} and P_d are the probabilities for energy transfer and deactivation of the excited chromophore, respectively:

$$P_{tr} = k_{tr}/\sum k \quad \text{and} \quad P_d = (k_r + k_{nr})/\sum k \qquad (23)$$

with $\sum k = k_r + k_{nr} + k_{tr}$.

The rate constants k_r, k_{nr}, and k_{tr} refer to radiative and nonradiative deactivation and energy transfer, respectively. Immediately after excitation of a chromophore within the micellar assembly, the probability for luminescence emission is given by

$$P_f^o = P_A \cdot P_f \qquad (24)$$

where $P_A = A/A^0$ is the fraction of chromophores that are not located in the immediate surroundings of the quencher where instantaneous quenching would occur, and $P_f = k_r/\sum k$ is the probability for luminescence emission. Let $P_f^{(m)}$ be the probability that luminescence occurs after m steps:

$$P_f^{(m)} = P_f P_A^{m+1} P_{tr}^m \qquad (25)$$

The total probability of fluorescence is

$$P_f^{total} = I/I^\circ = \sum_{m=0}^{m=\infty} P_A^{m+1} P_{tr}^m P_f = P_A \cdot P_f \cdot \frac{1}{1 - P_A P_{tr}} \qquad (26)$$

which can be written in the form

$$\frac{1}{P_f^{total}} = \frac{I_o}{I} = \frac{A^\circ}{A} \left[\frac{\sum k}{k_f} + \frac{k_{tr}}{k_f} \left(n\frac{Q}{A^\circ} - 1 \right) \right] \qquad (27)$$

where I_o and I are the emission intensities in the absence and presence of quencher, respectively.

A similar equation was employed by Shinitzky[26] to analyze the rate of energy transfer in micellar assemblies of sodium 2-hexadecylamino 6-naphthalene sulfonate (HNS$^-$) and N-palmitoyl L-tryptophan (PLT). In this case the quencher employed was N-hexadecyl pyridinium chloride. The rate constants for an energy transfer between adjacent fluorophores were determined as 7×10^8 and 5.5×10^8 sec^{-1} for HNS$^-$ and PLT micelles, respectively.

One problem with employing Equation 27 for this type of analysis is that it does not take into account the statistical nature of the distribution of quencher molecules over the host aggregates. Thus, instead of using the average occupation of a micelle by a quencher, the quantity Q in Equation 22, the probability of finding chromophores that are not in contact with quencher molecules is formulated more correctly by

$$P_A(i) = \frac{A^\circ - i \cdot n}{A^\circ} \tag{28}$$

Here, i indicates the number of quenchers per micelle, which is given by the Poisson distribution. The probability that a micelle with i quencher associations will fluoresce is

$$P_f(i) = P_A(i)P_F \cdot \frac{1}{1 - P_A(i)P_{tr}} \tag{29}$$

The ratio of luminescence quantum yields in the presence and absence of quencher is therefore given by

$$\phi/\phi^\circ = e^{-Q} \sum_{i=0}^{\infty} \frac{(Q)^i}{i!} \frac{P_A(i)}{1 - P_A(i)P_{tr}} \tag{30}$$

It should be noted that in the derivation of Equation 30 we have not considered a dipolar (Förster) type of energy-transfer mechanism. Therefore, it is strictly applicable only for short range interactions such as triplet energy transfer between identical chromophores. In many cases, in particular when excited singlet states are involved, excitation energy transfer can occur over a long range and does not require collisional contact of the reactants. In such a situation a different concept must be used for kinetic analysis, which will be discussed in the next section.

D. Long-Range Intramicellar Singlet Energy Transfer

In this section we shall deal with long-range energy transfer in micellar systems which does not require a collisional encounter of the reactants. Apart from the trivial case of radiative energy transfer implying emission and reabsorption of photons, a nonradiative mechanism can also be operative. Such a mechanism will become prominent when the emission spectrum of the donor and the absorption spectrum of the acceptor show a good overlap and, moreover, the electronic transitions accompanying the energy transfer are allowed. The latter condition is fulfilled in the case of singlet energy transfer:

$$^1D^* + {}^1A \rightarrow {}^1D + {}^1A^* \tag{31}$$

Consider again a donor-acceptor pair cosolubilized in the same micelle and separated by a distance R. The first-order constant for intramicellar energy transfer is given by the Förster expression:

$$k_t = \frac{1}{\tau} \left(\frac{R_0}{R}\right)^6 \tag{32}$$

where τ is the lifetime of the excited donor singlet in the absence of acceptor, and R_0 the critical distance for which $k_{tr} = 1/\tau$; that is, the energy-transfer rate and that for all other processes of deactivation are equal. The parameter R_0 can be calculated from

$$R_0^6 = \frac{9 \ln 10 \cdot \kappa^2 \cdot 10^{23}}{128\pi^5 n^4 N_A} \int_0^\infty F_D(\lambda)\epsilon_A(\lambda)\lambda^4 \, d\lambda \tag{33}$$

Here, κ^2 is a geometric factor depending on the relative position of the transition moments for donor emission and acceptor absorption, respectively. (The value of $\kappa^2 = 2/3$ if the molecular rotation is fast compared to the rate of energy transfer, and $\kappa^2 = 0.476$ if during the transfer of energy there is no molecular rotation. In both cases the two transition moment vectors are assumed to be in a random position with respect to each other.) In Equation 33, n is the refractive index of the solvent, and N_A Avogadros's number. $F_D(\lambda)$ is the spectral distribution of the donor luminescence which must be normalized such that

$$\Phi_O = \int_0^\infty F_D(\lambda) \, d\lambda \tag{34}$$

The absorption spectrum of the acceptor ^1A is expressed by the decadic molar extinction coefficient $\epsilon_A(\lambda)$. Equation 32 shows that the efficiency of nonradiative energy transfer depends strongly on the distance between the donor and acceptor. If both are solubilized in micellar solutions, the situation is such that the distance between different micelles largely exceeds R_O while the dimensions of one micellar aggregate are comparable to R_O. Therefore, one expects intramicellar energy transfer to proceed very rapidly while intermicellar transfer should be rather inefficient. As a consequence, the distribution of donor and acceptor molecules over the micelles will again play a primordial role in the kinetic analysis. Experimental results corroborating these predictions will be given below.

The probability of intramicellar singlet energy transfer is defined by

$$P_{tr} = \frac{k_{tr}}{k_{des} + k_{tr}} \tag{35}$$

where $k_{des} = 1/\tau$ represents the inverse fluorescence lifetime of the donor and hence is the sum of rate constants for radiative and radiationless deactivation observed in the absence of an acceptor. A combination of Equations 35 and 32 yields

$$p_t = \frac{1}{1 + \left(\dfrac{R}{R_O}\right)^6} \tag{36}$$

Consider now a donor-acceptor pair solubilized in the same host micelle. For simplification we assume the micelle to be spherical and D and A to be pointlike species. The distance R separating the two reactants is not fixed, and spatial averaging must be performed first. We shall discuss the case where both donor and acceptor molecules are associated with the micellar surface, e.g., cationic dyes in solutions of negatively charged micelles. Figure 8 illustrates this situation. We can express the distance R in terms of the micellar radius r_O and the angle gθ:

$$R^2 = 2 \, r_O^2 (1 - \cos \Theta) \tag{37}$$

The insertion of Equation 37 into Equation 36 yields

$$P_{tr} = \frac{1}{1 + \dfrac{8(1 - \cos \Theta)^3 \, r_O^6}{R_O^6}} \tag{38}$$

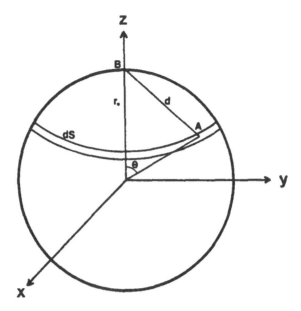

FIGURE 8.   Schematic illustration of a spherical micelle containing a donor (D) and acceptor in the surface region.

The probability of finding the acceptor at an angle Θ is given by

$$\frac{ds}{s} = \frac{2\pi r_O^2 \sin \Theta \, d\Theta}{4\pi r_O^2} = \frac{1}{2} \sin \Theta \, d\Theta \tag{39}$$

The average transfer probability is therefore

$$p_{tr} = \int_0^\pi p_{tr} \frac{ds}{s} = \frac{1}{2} \int_0^\pi \frac{\sin \Theta \, d\Theta}{1 + \frac{8r_O^6(1 - \cos)^3}{R_O^6}} = \frac{1}{2} \int_{-1}^1 \frac{dx}{1 + \frac{r_O^6 8(1 + x)^3}{R_O^6}} \tag{40}$$

Numerical integration yields the function displayed in Figure 9. From this curve one infers that the transfer probability reaches practically 100% when the critical distance for Förster-type transfer is three times greater than the micellar radius.[27]

The above considerations apply only to micelles occupied by one donor-acceptor pair. Under experimental conditions, there is a Poisson distribution of donors and acceptors over the host aggregates which must be accounted for. If one designates with $P_t(i)$ the probability of long-range energy transfer in micelles with i acceptor association, $P_t(i) = i \cdot P_t(1)$, one can express the quantum yield of donor fluorescence as

$$\Phi = \Phi_O \sum_{i=0}^\infty [1 - p_t(i)] \frac{\bar{n}_A^i}{i!} e^{-\bar{n}_A} \tag{41}$$

If one assumes that the probability for intramicellar transfer is unity for micelles with one donor and acceptor association, Equation 41 becomes

$$\Phi = \Phi_O e^{-\bar{n}_A} \tag{42}$$

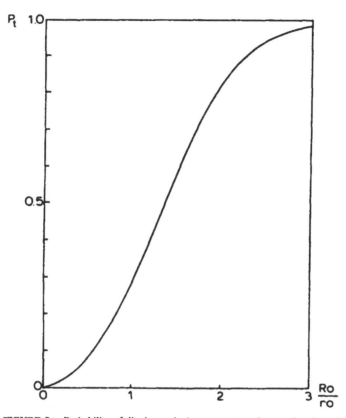

FIGURE 9. Probability of dipolar excitation energy transfer as a function of the ratio R_0/r_0 (R_0, critical distance for a Förster-type transfer: r_0, micellar radius). The curve was obtained by numerical integration of Equation 40 in Chapter 3.

These predictions have been verified meanwhile by several experimental investigations.[27-29] A particularly interesting case is that of the cyanine dyes, cyanine I:

and cyanine II:

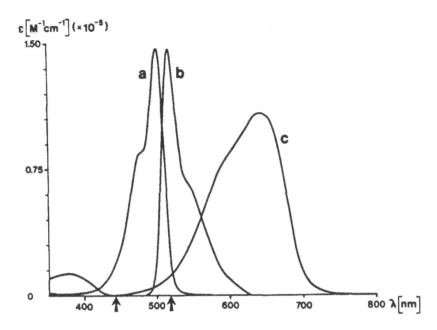

FIGURE 10.   Absorption and emission spectra of cyanine dyes employed for the study of intramicellar dipolar excitation energy transfer. (Cyanine I: a, absorption; b, fluorescence. Cyanine II: c, absorption.) *The arrows indicate the excitation and detection wavelengths selected for these experiments.*

Intramicellar singlet energy transfer was examined in micellar solutions of cetyl trioxy-ethylenesulfate, CTOES. These aggregates are negatively charged, which ensures absorption of the cationic dyes in the surface region. The fluorescence quantum yield of cyanine I in CTDES solution is high ($\Phi_o = 0.4$) and occurs in a wavelength region where cyanine II exhibits significant absorption (Figure 10). Thus, the requirements for efficient nonradiative energy transfer via the dipolar mechanism are fulfilled. From Equation 33 one obtains for the critical distance $R_o = 47$ Å. Since this corresponds to approximately three times the micellar radius, the probability of energy transfer should be practically 100%, even in aggregates that contain only one acceptor molecule (Figure 9). Therefore, Equation 42 should be applicable. Indeed, Figure 11 shows that a plot of ln (Φ/Φ_0) against acceptor concentration is linear, and the slope when evaluated according to Equation 3 yields an aggregation number of 93, which is in excellent agreement with $v = 96$, obtained by a different method. These results illustrate the extremely fast nature of intramicellar singlet energy transfer (Equation 38). Even for a fluorophore such as the cyanine dye I, whose fluorescence lifetime is only 2 nsec (Equation 31), the process occurs so rapidly that a single acceptor molecule per micelle suffices to totally quench the luminescence. These results are indeed related to the mode of action of the light-harvesting units in photosynthesis that conduct excitation energy from the Chl antenna molecules to the reaction centers in a very fast and efficient manner.

Ediger et al.[30] published several papers which elucidate the salient kinetic features of electronic excitation transport between chromophores via the resonance mechanism in organized molecular assemblies. These studies confirm the notion that infinite volume theory cannot be applied to such systems. The transport of single excitation energy along a spherical micellar surface was analyzed in detail. A statistical mechanical theory was developed, and the predictions from this model were compared to experimental results obtained with mixed micelles of Triton® X-100 and octadecyl rhodamine B. The latter was excited by a picosecond laser pulse, and the time course of intramicellar energy transfer between different rhodamine B molecules was followed by monitoring the temporal decay of the fluorescence

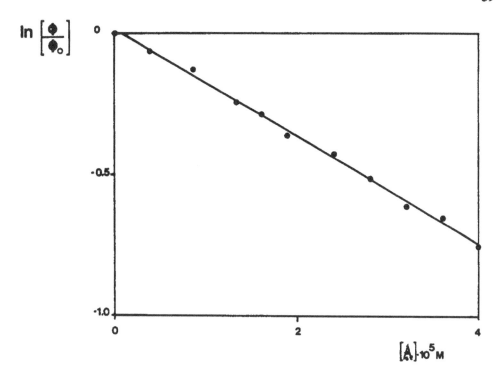

FIGURE 11. Plot of ln ϕ/ϕ^0 as a function of the acceptor concentration (cyanine II). The donor is cyanine I at $6.26.10^{-6} \, M$ concentration in CTOES micelles [CTOES] = $5.10^{-3} \, M$.[27]

polarization. In the analysis of the data, the assumption was made that the distribution of the chromophore over the micelles follows a Poisson law, i.e., Equation 2. The orientation-averaged Förster radius for rhodamine B corresponds to 51.5 Å. The surfactant derivative has a slightly lower R_0 value, 48 Å, since it is oriented at the micellar surface. Using this Förster radius and a size of 74 Å for the Triton® X-100 micelles, very good agreement between theory and experimental results was obtained. The assembly of the fluorophore on the micellar surface was found to render resonance energy transfer particularly effective. Thus, at an average occupancy of 2.2 octadecyl rhodamine B molecules per micelle, the probability that the excitation is on the molecule which originally absorbed the light decreases to 50% within a time delay of only 5 nsec.[30]

III. PHOTOINDUCED ELECTRON TRANSFER IN MICELLAR ASSEMBLIES

The study of light-induced electron-transfer reactions in micellar assemblies has become a topic of intensive investigations. Advantage is taken of the distinct microenvironment present in these systems to model electron conduction and interfacial charge transfer processes occurring in biological membranes. Another aim of this work is the design of functional molecular units to optimize a light-induced charge separation. Progress in this domain of artificial photosynthesis has been very rapid, and some of the most important findings will now be discussed. We shall first address the question of the electrical potential distribution in the vicinity of charged molecular assemblies. These local fields play an important role in the kinetics of interfacial redox reactions, in particular the retardation of the back reaction following light-inducted electron transfer.

A. Electrical Potential of a Solution in the Vicinity of a Charged Molecular Assembly

A qualitative description of the potential distribution in an electrolyte surrounding a spherically shaped aggregate such as a micelle or a colloidal semiconductor is given in

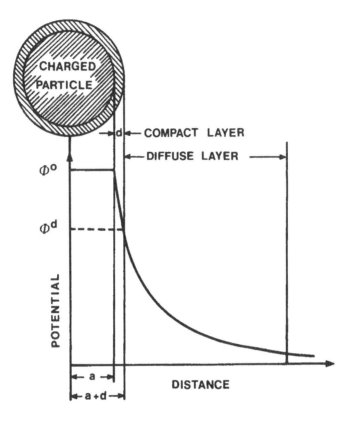

FIGURE 12.   Ionic double layer and electrical potential gradient surrounding
a charged spherical particle in solution.

Figure 12. At the surface of the particle the potential is ϕ_o. The solution layer surrounding
the particle is divided into two parts: the Stern-Graham or compact layer and the Gouy-
Chapman or diffuse layer. The width of the compact layer corresponds to the distance of
closest approach of the counterions, i.e., 1 to 5 Å depending on whether the adsorbed ions
are solvated or not. Within this region the potential decreases in an approximately linear
fashion, reaching the value ϕ^d in the outer Helmholtz plane. The diffuse layer extends from
there into the solution and has a much larger width, which can attain several hundred
nanometers. The potential distance function with this layer is commonly derived from the
Poisson-Boltzmann equation:[31]

$$\Delta\phi = -\frac{1}{\epsilon_o\epsilon}\sum_i z_i e n_i^b \exp(-z_i e\phi/kT) \tag{43}$$

where Δ is the Laplace operator. For a planar interface such as a charged monolayer or
membrane, Δ has the form

$$\Delta = \delta^2/\delta x^2 + \delta^2/\delta y^2 + \delta^2/\delta z^2 \tag{44}$$

while for a spherical particle the Laplacian becomes:

$$\Delta = \frac{1}{r^2}\frac{d}{dr}\left(r^2\frac{d}{dr}\right) \tag{45}$$

In Equation 43, Σ_O is the permeativity of the free space; Σ is the dielectric constant or relative permeativity of the solution; z_i and n_i^b are the charge and the bulk concentration, respectively, of the ion i; and e is the elementary charge. The solution of the Poisson-Boltzmann equation is straightforward for the case of a flat surface, where the Laplacian reduces to d^2/dx^2. If the solution contains a symmetric, i.e., 1:1 ($z_+ = z_- = z$), electrolyte, integration from some point in the bulk where $\phi = 0$ and $d\phi/dx = 0$ up to some point in the diffuse layer [x > (d + a)] yields for the potential gradient

$$\frac{d\phi}{dx} = -\frac{2\kappa kT}{2e} \sin h(ze\phi/kT) \tag{46}$$

where

$$\kappa = (e^2 \sum_i n_i^b z_i^2 / \epsilon_O \epsilon kT)^{1/2} \tag{47}$$

is the reciprocal Debye length. In water at 25°C the value of κ is given by $\kappa = 3.28\sqrt{I}$ nm^{-1}. I is the ionic strength, $(^1/_2 \Sigma_i c_i) z_i^2$, where c_i is the ionic concentration in (moles per liter). In 10^{-3} M NaCl, $1/\kappa = 9.6$ nm; and for the systems of interest in colloidal science, $1/\kappa$ ranges from a fraction of a nanometer to about 100 nm. Integration of Equation 45 gives the potential distance function in the diffuse layer:

$$\tanh(ze\phi/4kT) = \tanh(ze\phi^d/4kT) \exp[-\kappa(x - d)] \tag{48}$$

For low potentials (zeΦ < 4kT) the substitution tanhy $\simeq$ y can be made (Debye-Hückel approximation), and Equation 48 reduces to

$$\phi = \phi^d \exp[-\kappa(x - d)] \tag{49}$$

indicating an exponential decrease of the potential with distance. For charged molecular assemblies, such as surfactant micelles or colloidal semiconductors, the Laplace operator in the Poisson-Boltzmann equation is expressed in spherical coordinates (Equation 45). Unfortunately, in this case it is impossible to solve the equation analytically. Recourse to the Debye-Hückel approximation gives the following distance dependence of the electrostatic potential:

$$\phi = \phi^d \frac{(a + d)}{\tau} \exp[-\kappa(\tau - a - d)] \tag{50}$$

where d is the distance of closest approach of the ions to the colloid surface. It should be noted that the Debye-Hückel approximation cannot be used to evaluate the electrostatic potential in the vicinity of micelles or vesicles. This is due to the high surface potential of these aggregates. Typically, the diffuse layer potential for micelles and vesicles is between 50 and 200 mV.[32] Similar values are found for inorganic colloids. Thus, $\phi e^d/kT$ is in the range of 2 to 8 at room temperature, which may be too high to render the Debye-Hückel model applicable. It should be noted, however, that although Equation 50 would not give the potential in the immediate vicinity of such colloidal particles, it is still applicable at distances farther from the surface where ϕ is low and the condition eϕ/kT < 4 is fulfilled.

For an adequate description of the potential within the entire range of the diffuse layer, one must perform a numerical integration of the Poisson-Boltzmann equation. Extensive work has been published on this subject and has recently been reviewed.[31] Solutions for a

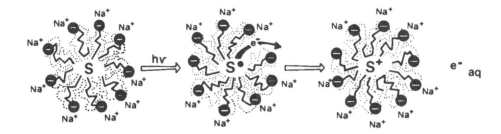

FIGURE 13.   Schematic illustration of the photoionization of a solubilizate molecule S in an anionic micelle.

large variety of experimental conditions are now available. Much of the original work has been concerned with the calculation of the electrical field around a single aggregate. These earlier calculations neglect the mutual penetration of the diffuse layers of adjacent particles which occurs frequently since there are always many colloidal particles present in solution. This is untenable for micellar solutions at low ionic strength, where the diffuse layers overlap even if the concentration of aggregates is as small as 10^{-5} M. As a result of this neglect, the potential difference between the surface of the micelle and the solution is overestimated. More recently, cell models have been applied which take the overlap of diffuse layers from adjacent aggregates into consideration.[33]

Having established a quantitative description of the electrostatic potential profile in the vicinity of charged colloidal particles, we can now proceed to analyze the role which these local fields play in redox processes that occur in micellar assemblies. We shall restrict ourselves to the domain of photodriven electron-transfer processes. Extensive literature exists concerning thermal redox reactions in such systems, which are covered by a review.[34] The following examples illustrate how advantage can be taken of the electrostatic microenvironment of ionic micelles to achieve light-induced charge separation and mimic in this fashion the primary act of photosynthesis.

B. Photoionization

Photoionization is the simplest type of interfacial electron transfer in micellar assemblies. A schematic illustration of such a process is given in Figure 13. The photoactive probe is incorporated in the micellar interior, and upon photoexcitation it injects an electron into the surrounding aqueous phase, where it becomes hydrated. The first studies of photoionization reactions in micellar assemblies were carried out with pyrene[36,37] as a sensitizer which was solubilized in NaLS micelles. A mode-locked, frequency-doubled ruby laser was used as a light source. The photoionization was found to occur via a biphotonic mechanism: the first 347-nm laser photon-excited pyrene to the S_1 state. This absorbed a second photon to produce a highly excited singlet state from which the electron was injected into the water.

A very important discovery was made during these studies concerning the back reaction of hydrated electrons with the pyrene parent cation. This was found to be impaired by anionic micelles, while it occurred readily in nonionic or cationic aggregates.

The inhibition of back electron transfer can be attributed to the negative charge of the NaLS micelles, which produces an electrostatic barrier for the reentry of the photoejected electrons in the micellar interior. The surface potential of an NaLS micelle is -150 mV.[35] Assuming a Boltzmann distribution, the probability of a hydrated electron acceding to the surface of NaLS micelles is 2.5×10^{-3}. The rate of the back reaction is retarded by about the same factor.

These findings were subsequently confirmed by experiments involving monophotonic photoionization of a variety of solubilized substrates.[13,14,38,39] For example, sunlight irradiation of N,N,N',N'-tetramethyl benzidine (TMB) dissolved in aqueous solutions of NaLS micelles produces TMB^+ cations and hydrated electrons:

$$TMB \rightarrow TMB^+ + e_{aq}^- \qquad (51)$$

Since the back reaction between TMB^+ and e_{aq}^- is prevented by the negative surface charge of the anionic aggregates, the hydrated electrons dimerize in the aqueous phase to produce hydrogen:

$$2e_{aq}^- \rightarrow 2OH^- + H_2 \qquad (52)$$

and the overall reaction induced by sunlight is

$$TMB \rightarrow TMB^+ + \tfrac{1}{2}H_2 + OH^- \qquad (53)$$

This process stores about 0.8 eV in free energy per absorbed photon.

A very interesting study of the effect of micellar charge on photoionization efficiency was carried out by Grand et al.[40] Perylene, a hydrophobic aromatic molecule, was solubilized in the interior of the micelles, and the quantum yield of electron ejection in the aqueous phase was determined as a function of excitation wavelength. The ionization threshold was found to be around 270 nm (4.6 eV) for anionic, nonionic, and cationic micelles. There appeared to be no significant effect of the micellar surface potential on the photoionization threshold. This is perhaps not surprising since the potential difference between the micellar and aqueous phase made only a small contribution to the total ionization energy in the cases examined. A parameter which is much more sensitive to the interfacial potential is the yield of ionization. The anionic aggregates exhibited a much steeper increase in the photoionization curve above threshold, presumably due to a decrease in geminate recombination.

Photoionization thresholds have also been determined for other solubilized molecules. For example, the values found for TMB and Chl a are 3.4[41] and 4.3 eV,[44] respectively. Bernas et al.[44] have used the relation

$$I_{ph} = I_g + P_+ + V_0 \qquad (54)$$

to predict the ionization threshold energy (I_{ph}) for solubilized substrates. In Equation 54, I_g is the gas phase ionization potential, V_0, -1.2 eV, is the energy of a thermalized electron in water with respect to vacuum, and P_+ is the polarization energy of the cation. The latter was calculated using the Born expression:

$$P_+ = -\frac{e^2}{R}\left(1 - \frac{1}{D_{op}}\right) \qquad (55)$$

where R is the effective ionic radius of the cation, and D_{op} is the optical dielectric constant of the medium which surrounds it. The agreement between predicted and observed I_{ph} values was satisfactory. Ohta and Kevan[42] have employed a more complex expression for the total polarization energy P_+ to account for differences of photoionization thresholds observed in solutions of micelles and vesicles.

It is worthwhile to consider in more detail the mechanism of a photoionization process in micellar solutions. This is an electron-transfer reaction where the donor, i.e., the excited chromophore, and the acceptor, i.e., the aqueous phase, are locally separated. Therefore, a photoionization reaction is likely to be in the nonadiabatic regime. The rate of the tunneling transition of the electron across the micelle-water interface will depend critically on the distance between the solubilized donor and the first layer of water molecules at the micellar surface. In order to obtain reasonable yields for electron ejection, this tunneling transition must be very fast. Photoionization frequently involves an electronically and vibrationally

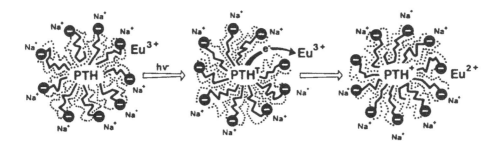

FIGURE 14.   Interfacial electron transfer from triplet-state phenothiazine to a Eu^{3+} ion located in the micellar Stern layer.

excited singlet state whose lifetime is of the order of 10^{-11} to 10^{-12} sec. The interfacial electron emission must occur at a similar rate in order to compete with vibrational relaxation. Thus, only solubilizate molecules residing close to the micelle water interface should be able to undergo efficient photoionization.

The low-temperature EPR studies of Kevan et al.[44] with frozen aqueous solutions of micelles and vesicles confirm these predictions. For example, the primary photoionization efficiency of TMB at 77°K is higher in cationic than in anionic micelles since, in the former type of aggregates, the solubilization site of TMB is closer to the interface than in the latter. (Recall, however, that the net charge separation efficiency at room temperature is much higher in anionic as compared to cationic micelles since in the negatively charged aggregates the back reaction of electrons with TMB^+ is impaired by the electrostatic barrier present at the surface.) The studies of Kevan et al. have shown that the photoionization yield in NaLS solution can be improved by exchanging the sodium with tetramethylammonium ions. The latter enhance the water penetration in the palisade layer, shortening the tunneling distance of the electron to the aqueous phase.

Photoionization processes have also been investigated in the functional micellar assemblies where the photoactive chromophore is chemically attached to the surfactant chain. A particularly intriguing example is that of the amphiphilic phenothiazine derivative sodium-12-(10-phenothiazinyl)dodecylsulfonate.[41] Here, monophotonic photoionization is only observed when the surfactants are present in micellar form, i.e., at concentrations above the CMC. Self-assembly drastically enhances the efficiency of charge separation and the lifetime of the phenothiazine radical ion. This provides a very nice example of the importance of molecular self-organization in determining the features of photoionization processes.

C. Intramicellar Electron Transfer

In 1975 we carried out the first time-resolved experiment on light-induced intramicellar electron-transfer reactions.[48] Phenothiazine was incorporated into SDS micelles and served as the photoactive electron donor, while Eu^{3+}, present in the Stern layer of the aggregates, was the electron acceptor (Figure 14). The dynamics of the reaction

$$PTH(T) + Eu^{3+} \rightarrow PTH^+ + Eu^{2+} \tag{56}$$

where PTH(T) represents the triplet state of phenothiazine, were investigated by the laser photolysis technique. The decay was completed within a few hundred nanoseconds and could not be fitted to a conventional first- or second-order rate equation. It was soon recognized that the rate laws derived in Section II.A for intramicellar triplet energy transfer were also applicable to this type of redox process.

The first case analyzed in detail in this manner was the photoinduced reduction of $Ru(bipy)_3^{2+}$ by MPTH:[49]

$$*Ru(bipy)_3^{2+} + MPTH \rightarrow Ru(bipy)_3^{+} + MPTH^{+} \qquad (57)$$

where $*Ru(bipy)_3^{2+}$ designates the charge transfer excited state of the ruthenium complex. The temporal behavior of the concentration of this excited state after laser excitation is described by

$$[*Ru(bipy)_3^{2+}] = [Ru(bipy)_3^{2+}] \, (t = 0) \, exp(-k_d t + \bar{n}_A[exp(-k_q t) - 1]) \qquad (58)$$

Here k_d is the rate constant for the deactivation of excited states in the absence of MPTH, and $\bar{n}_A$ corresponds to the average occupancy of a micelle by MPTH molecules. The kinetic evaluation carried out for tetradecyl-trioxyethylene sulfate micelles yields for the specific rate of electron transfer between one donor-acceptor pair a value of $k_q = 6 \times 10^6 \, sec^{-1}$, and for the micellar aggregation number $v = 95$ surfactant molecules per micelle. This implies that the average time required for electron transfer from the MPTH molecule to electronically excited $Ru(bipy)_3^{2+}$ is 170 nsec. Similar values have been obtained for a number of other donor-acceptor pairs.[50,51] Thus the cross sections observed for light-induced redox reactions in micellar assemblies are comparable to those for intramicellar triplet energy transfer discussed in Section II.A. They do not significantly exceed the limit expected for a diffusion controlled process. The implication of this finding is that long-range electron tunneling does not make an important contribution for the cases studied so far. These investigations were all carried out at room temperature. It would be interesting to perform low-temperature experiments in order to render tunneling effects easier to discern. Note that there is a difference between intramicellar electron transfer and photoionization in that, in the latter case, the reactive excited state is so short-lived that there is no possibility of diffusional displacement during the photoemission of electrons. If the chromophore which undergoes photoionization is residing in the hydrocarbon-like interior of the micelles, the only possibility for electron transfer to the aqueous phase is via long-range tunneling. Returning now to Figure 14, we are interested in exploring the fate of the products of the electron-transfer reaction, i.e., the Eu^{2+} and PTH^{+} cations. Since both species remain associated with the host aggregates, there will be a rapid back reaction:

$$Eu^{2+} + PTH^{+} \xrightarrow{k_b} Eu^{3+} + PTH \qquad (59)$$

Suppose that the experimental conditions chosen are such that no more than one pair of product ions per micelle is produced during the light reaction. In such a case, back reaction follows a simple first-order rate law. Experimentally, one finds that k_b is on the order of 2×10^6 to $10^7 \, sec^{-1}$. This is the expected result, since it corresponds to the diffusion-controlled limit for the back electron-transfer rate.

In conclusion, the kinetic consequences arising from the sequestering of the donor and acceptor molecule in a micelle are dramatic. In homogeneous solution both forward and backward electron transfer would follow a second-order rate law. The reaction order is reduced to 1 if the reactants are confined to the micellar space and only one pair is involved in the redox process. Furthermore, through compartmentation, the reactants are brought into close proximity, which enhances the electron-transfer event.

Unfortunately, in the examples discussed so far, both the forward and backward electron transfers were accelerated. As a consequence, the time period over which light-induced charge separation is sustained extends over at most a few hundred nanoseconds. This is hardly enough for coupling the photoredox process with suitable fuel generating dark reactions such as water oxidation and reduction. A dramatic progress in sustaining charge separation is achieved by again exploiting the local electrostatic potential present at the micelle-water interface (Figure 12) and by inhibiting in this way the thermal back reaction.

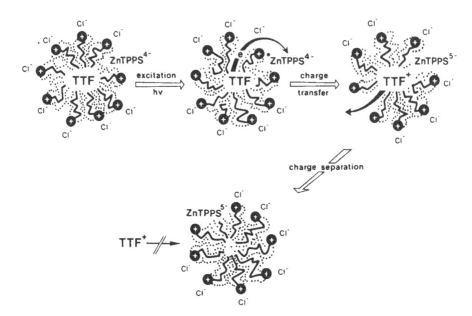

FIGURE 15.   The TTF-porphyrin system as an example of light-induced charge separation by ionic micellar assemblies. The porphyrin triplet state serves as an acceptor of the electron, and TTF serves as a donor.

Consider, for example, a situation where the electron donor is tetrathiafulvalene (TTF) and the acceptor is zinc tetraphenylporphyrin tetrasulfonate, $ZnTPPS^{4-}$, (Figure 15). The latter is excited by visible light, which promotes electron transfer from TTF to $ZnTPPS^{4-}$. The radical ion pair TTF + -$ZnTPPS^{5-}$ is thereby produced within the aggregate. In a cationic micelle, TTF^{+} is clearly destabilized with respect to the aqueous bulk solution. Therefore, once it reaches the surface it will be ejected into the water. Conversely, the $ZnTPPS^{5-}$ radical is electrostatically more stable in the micellar than in the aqueous phase and will thus remain associated with the surfactant aggregate.

Once A^{-} and D^{+} are separated, the diffusional reencounter is obstructed by the ultrathin barrier of the micellar double layer. Figure 15 illustrates the successful charge separation in the $ZnTPPS^{4-}$-TTF system by cationic CTAC. The porphyrin was excited by a 530-nm laser pulse and the decay of the triplet state and formation of $ZnTPPS^{5-}$ were monitored at 840 and 893 nm, respectively.[52] The forward reaction follows the multiexponential time law given by Equation 9, and from the kinetic analysis one obtains $k_q = 7 \times 10^4 \ sec^{-1}$. Through subsequent ejection of TTF^{+} from the CTAC micelles, the back reaction is impaired, the decay of $ZnTPPS^{5-}$ being hardly visible on a 500-μsec time scale.

In striking contrast to these results, one finds in homogeneous solutions, e.g., methanol, no formation of redox products from the reductive quenching of triplet $ZnTPPS^{4-}$ by TTF. Practically all radical ion pairs (TTF^{+}-$ZnTPPS^{5-}$) apparently undergo back reaction in the solvent cage. This provides an illustrative example of how, by using molecular assemblies, light-induced charge separation can be achieved.

In general, the efficiency of the charge-separation process will depend crucially on the relative rates of ejection and intramicellar back transfer of electrons from A^{-} to D^{+}. The latter reaction is thermodynamically favorable and can in principle occur very rapidly. The rate of ejection of an ion from a charged micelle, on the other hand, is expected to depend critically on the degree of hydrophobic interaction of A^{-} with the aggregated surfactant molecules.

Using this principle, light-induced charge separation has been achieved for a number of

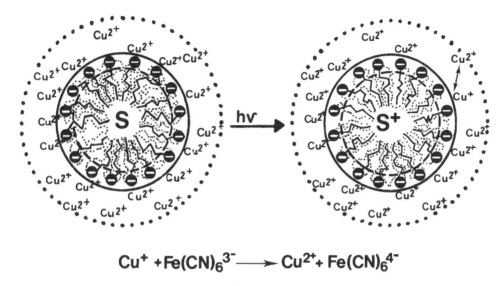

$$Cu^+ + Fe(CN)_6^{3-} \longrightarrow Cu^{2+} + Fe(CN)_6^{4-}$$

FIGURE 16.   Photoinduced electron transfer in an anionic surfactant micelle with a transition metal counterion atmosphere.

donor-acceptor pairs. Noteworthy are the examples where pyrene served as an electron acceptor and dimethyl aniline as a donor[51,52] as well as the photoinduced reduction of duroquinone (DQ) by Chl a.[53]

The laser photolysis analysis of the latter system showed that a large fraction of Chl a^+ does not undergo back reaction with the reduced quinone (DQ^-) due to efficient ejection of the latter from the micellar into the aqueous phase. The rate of DQ^- ejection was estimated to be 10^8 sec^{-1}. Since the intramicellar back reaction typically occurs at a rate 10 to 20 times slower, one infers that the DQ^- escape yield from the micelle is at least 80%. As the reentry of DQ^- into the surfactant assembly is prevented by the micellar surface charge, the lifetime of Chl a^+ is prolonged by several orders of magnitude with respect to homogeneous solutions.

D. Photoinduced Electron Transfer in Functionalized Micellar Assemblies

Further progress in the development of molecular assemblies capable of light-induced charge separation was made by designing and synthesizing surfactants with suitable functionality. These are distinguished from simple surfactants by the fact that part of the micelle participates in the redox events. The role of the conventional micelle in a photoredox reaction, as described in the previous section, was to provide solubilization sites for hydrophobic reactants and to assist in light-induced charge separation. Functionalized micelles have the additional advantage of displaying cooperative effects which will be illustrated in the examples below. A simple way to accomplish functionalization is to replace the inert counterion atmosphere of an ionic micelle with a reactive one. Counterions were chosen which could act as electron acceptors.[54,55]

Consider, for example, the case of a transition metal ion micelle which can be obtained by replacing the Na^+ in SDS by Cu^{2+}.[48] The electron transfer from a solubilized sensitizer to the counterion atmosphere is illustrated in Figure 16. An illustrative case is that where D = N,N-dimethyl-5,11-dihydroindolo-3,3,6-carbazole (DI). When dissolved in SDS micelles, DI displays an intense fluorescence, and the fluorescence lifetime measured by laser techniques is 144 nsec. Introduction of Cu^{2+} as a counterion atmosphere induces a 30-fold decrease in the fluorescence yield and lifetime of DI. The detailed laser analysis of this system showed that in $Cu(DS)_2$ micelles there is an extremely rapid electron transfer from the excited singlet to the Cu^{2+}. This process occurs within an average time of 4 nsec, and hence can compete efficiently with fluorescence and intersystem crossing. This astonishing result must be attributed to a pronounced micellar enhancement of the rate of the transfer reaction. It is, of course, a consequence of the fact that within such a functional surfactant unit, regions with extremely high local concentrations of Cu^{2+} prevail. (Theoretical estimates predict the counterion concentration in the micellar Stern layer to be between 3 and 6 M.) The previously developed theoretical approach for intramicellar redox reactions[55] can explain satisfactorily such a behavior. In fact, the case where an excited species, S*, distributed initially in a random fashion within a surfactant assembly, diffuses to a reactive interface can be treated mathematically in a closed fashion. Fick's equation:

$$\frac{\partial [S^*](r,\, t)}{\partial t} = D \nabla^2 [S^*](r,\, t) \tag{60}$$

must be solved subject to the following three boundary conditions:

$$[S^*](r_0,\, t) = 0 \quad \text{all} \quad t \tag{61}$$

$$[S^*](r,\, 0) = [S^*]_0 \quad \text{all} \quad r \leqslant r_0 \tag{62}$$

$$\lim r \cdot [S^*](r,\, t) \to 0 \quad \text{all} \quad t \tag{63}$$

In Equation 60 $[S^*]$ (r,t) is the ensemble-averaged concentration of the excited sensitizer inside the micelle at time t and at a distance r from the center of the micelle; D is the diffusion coefficent; and ∇^2 is the Laplacian operator. The micellar radius is represented by r_0. The boundary condition in Equation 61 expresses the fact that the surface of the micelle is constituted by reactive transition metal ions and hence may be regarded as a perfectly absorbing boundary for S*. The second condition, in Equation 62, specifies that at time t = 0 all interior positions of the micelle are accessible to the confined, diffusing species; that is no localization of the confined species is assumed at t = 0. The limit specified by Equation 63 is a finiteness condition which formalizes the physical constraint that the reactant concentration remains finite for all times and all interior points of the assembly. The integration of Equation 60 first gives the concentration of the sensitizer as a function of both radial position and time. After spatial averaging one obtains for the experimentally measured quantity.

$$[S^*](t) = \frac{6[S^*](t_0)}{\pi^2} \sum_{n=1}^{\infty} \frac{1}{n^2} \exp \left\{ \frac{-n^2 \pi^2 D}{r_0^2} t \right\} \tag{64}$$

Given the exponential structure of this equation, on a time scale for which $t > r_0^2/\Pi^2 D$, the n = 1 term in the summation will dominate the kinetic behavior of the system. Since the ratio $r_0^2/\Pi^2 D$ for micellar aggregates is typically on the order of a few nanoseconds, a necessary consequence of Equation 64 is that kinetic processes which take place on a time scale greater than 1 nsec should exhibit essentially first-order kinetics. This is indeed observed in the case of the luminescence decay of DI in the Cu^{2+} counterion micelles described above. From Equation 64 a fluorescence lifetime of 4 nsec, i.e., a rate constant for electron transfer from DI* to Cu^{2+} of 2.5×10^8 sec^{-1}, is compatible with a micellar radius of 15 Å and a diffusion coefficient of 5×10^{-7} cm^2/sec. It is important to note that the rate constant observed in this functionalized micelle is about 50 times greater than that for electron transfer between a donor-acceptor pair in a simple surfactant aggregate. We shall return to the kinetic problem of charge carrier trapping by reactive surfaces in Chapter 3 when we discuss electron and hole reactions at the surface of colloidal semiconductor particles.

The significance of the functional organization in Figure 16 also becomes evident when the back reaction of Cu^+ and DI^+ is considered. Previous studies have shown that the intramicellar electron transfer from Cu^+ to DI^+, though thermodynamically highly favorable, cannot complete kinetically with the escape of Cu^+ from the native micelle into the aqueous phase. An efficient escape route is provided by the exchange with one of the Cu^{2+} present in high local concentration in the Gouy-Chapman layer. By mere electrostatic arguments, the latter ion is exp ($e\phi/kT$) times more likely to be absorbed on the micellar surface than Cu^+. Once in the aqueous phase, Cu^+ can undergo a second redox process, such as the reduction of $Fe(CN)_6^{3-}$. The back reaction of $Fe(CN)_6^{4-}$ with the oxidized donor, DI^+, is prevented by the negatively charged micellar surface. Hence, such a system is successful in storing light energy originally converted into chemical energy during the photoredox process.

Over the last years sophisticated functionalized surfactants have been developed that show unique and highly promising effects in light-induced charge separation.[56,57] For example, crown ether-type amphiphiles complexed with suitable transition metal ions have been successfully employed as molecular electron-storage devices.[58,59] Electron-transfer processes in such aggregates follow rules similar to those followed by the reactions discussed above and hence will not be treated in more detail here.

E. Light-Induced Charge Separation Through Hydrophobic Electron Storage in Micellar Aggregates

Light-induced charge separation and electron storage are very important aspects of photon energy conversion. Recall the function of the plastoquinones in the photosynthetic membrane. These amphiphylic molecules act as terminal electron acceptors in photosystem II. The plastoquinone pool serves as an electron reservoir linking the two photosystems and thereby enabling the operation in series. It has been possible by using suitable molecular engineering to mimic electron-storage effects in artificial devices. The strategy is to use, in a judicious way, the hydrophobic and electrostatic interactions present in solutions of organized assemblies such as micelles and vesicles. Consider, for example, the case where a water-soluble sensitizer (S), after light excitation, reduces a viologen carrying a hydrocarbon tail, such as C_nMV^{2+}:

The redox reaction is carried out in a solution containing positively charged micelles (Figure 17). The length of the hydrocarbon tail has been chosen such that the oxidized form of the viologen is present mainly in water while the reduced one shows a strong affinity for the micellar phase. In other words, the strategy for achieving charge separation is to use a water-soluble electron acceptor which upon reduction turns very hydrophobic and hence is rapidly sequestered into the micelles. Since the micelle is positively charged, the sensitizer cation is rejected from the surface by electrostatic repulsion, and this effect prevents the thermal back electron transfer. In the case of $C_{14}MV^{2+}$, the equilibrium constant for asso-

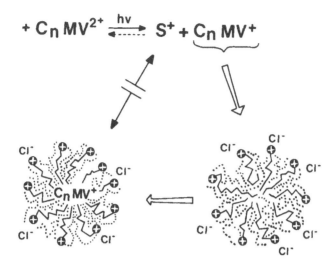

FIGURE 17.   Principle of hydrophobic charge storage following photoinduced electron transfer in an organized molecular assembly.

ciation with CTAB micelles is $1 \times 10^3 M^{-1}$, while that of the reduced form is much higher, $2.7 \times 10^5 M^{-1}$. As a consequence, in a 0.01 M CTAB solution, the fraction of $C_{14}MV^{2+}$ residing inside the micelle is 9%, while that of $C_{14}MV^+$ is 96%. Electron storage and charge separation in this system have been studied using $Ru(bipy)_3^{2+}$ or zinc tetrakis (N-methylpyridyl) porphyrin as the sensitizer.[60-65] A 400-fold retardation in the rate of the back electron transfer as compared to homogeneous solutions was observed. From the kinetic analysis, the rate constant for association of $C_{14}MV^+$ with the CTAB micelles was found to be 5×10^8 $mol^{-1} sec^{-1}$ while the exit rate from the micellar aggregate was $2 \times 10^3 sec^{+1}$. This is a particularly nice example of how the dynamics of photoredox events can be affected by microheterogeneous systems.

The charge separation effect is not restricted to viologens as electron carriers. For example, we have also been able to obtain charge storage by using decamethyl ferrocenium ions as electron carriers. In this case, there is again a drastic increase of the hydrophobicity of the acceptor upon reduction which can be exploited to trap it in hydrophobic regions of micellar or vesicular aggregates, protected from the access of the oxidized sensitizer by an electrostatic barrier.

F. Micellar Model Systems for Photosynthesis

To conclude this section on electron-transfer reaction in micellar assemblies, we briefly discuss systems which imitate photosystem I of plants photosynthesis. An instructive example is the Krasnovsky reaction, in which Chl a acts as a photosensitizer for transfer of an electron from a donor, such as NADH or phenylhydrazine, to an acceptor, such as methylviologen, under the action of red light. In the presence of an enzyme, e.g., hydrogenase, the reduced electron acceptor produces hydrogen from water:

In Krasnovsky's experiments[66] the chlorophyll was solubilized in Triton® x-100 micelles. The detailed mechanism of the Krasnovsky reaction was investigated by Kalyanasundaram and Porter[67] using flash photolysis. It was shown that oxidative quenching of the Chl a triplet state was operative. The spatial separation of the products of the photoredox reaction and the statistical distribution of the sensitizer among the micelles was found to have a considerable influence on the overall efficiency of the system. The Krasnovsky reaction served as a prototype for hundreds of similar three-component systems using a wide variety of sensitizers and electron relays. Hydrogen was produced at the expense of a donor molecule such as cysteine, EDTA, or triethanolamine.[68] The investigation of such sacrificial systems has been spurned in particular by the introduction of colloidal metals instead of hydrogenase as the water reduction catalyst.

IV. LIGHT-INDUCED ELECTRON TRANSFER IN INVERTED MICELLES AND MICROEMULSIONS

Inverted micelles are present in solutions of various amphiphiles in nonpolar organic solvents.[69] The structure is such that the polar head groups of the surfactant molecules constitute the core of the aggregate while the hydrophobic tails extend into the surrounding solution. Among the amphiphiles capable of forming inverted micelles, diisooctyl sodium sulfosuccinate (Aerosol OT, AOT) has received particular attention. The interest focuses on the ability of AOT micelles to solubilize relatively large amounts of water in a variety of hydrocarbon liquids. The water is accommodated in the polar center of the aggregates, where it forms spherical pools the sizes of which are controlled by the AOT-water ratio (w = $n(AOT)/n(H_2O)$. The aqueous core is surrounded by a monolayer of AOT molecules resulting in the configuration shown in Figure 18. The total micellar radius R is composed of the length of an AOT molecule (~ 11 Å) and the radius of the water pool. With isooctane as a solvent the latter can be estimated from the relation:

$$r(Å) = 1.5w \tag{65}$$

The water clusters solubilized in inverted micelles provide a molecular organization similar to the pockets of water in bioaggregates [70,71] such as biomembranes or mitochondria.[72] The inverted micelles provide suitable model systems for exploring the salient features of electron-transfer reactions in such aqueous microphases surrounded by a hydrophobic environment. Research activity in this field has been particularly active during recent years, and a few illustrative examples of such studies will be presented here.

Willner et al.[73] have performed pioneering studies on light-induced electron-transfer processes in inverted micelles. The sensitizer, e.g., Ru(bipy)$_3^{2+}$ or a porphyrin was dissolved in the water pool together with a sacrificial electron donor. An amphiphilic viologen derivative, such as 1.1'-dihexadecyl 4,4'bipyridinium chloride, was used as an electron relay. As was pointed out in the previous paragraph, the characteristic feature of this class of acceptors is that the hydrophobicity increases strongly upon reduction. In the reduced state, the relay moves to the bulk oil phase, leading to a charge separation.[74]

Atik and Thomas[75] and Brochette et al.[76] have investigated the kinetic features of electron-transfer reactions in inverted micelles by applying laser photolysis techniques. The compartmentation of the reactants in the water pool leads to rate laws equivalent to those observed for simple micelles. Thus, Equation 58 was found to be directly applicable to describing the time dependence of the concentration of excited sensitizer. This shows that the distribution of the quencher over the water pools obeys Poisson statistics as was observed for simple micelles. These studies have also given valuable information on the rate of migration of a variety of substrates between different water pools.

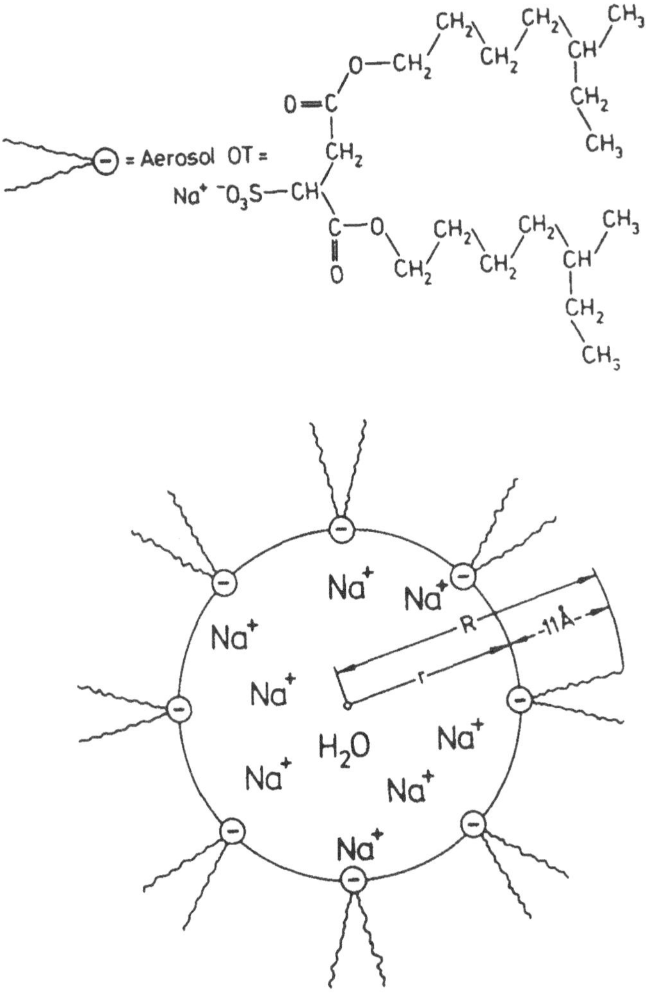

FIGURE 18.   Illustration of an inverted micelle with an aqueous core.

Ulrich and Steiner[77] have investigated the effect of micellar size and magnetic field on the charge recombination involving radical ions in inverted micelles of cetyltrimethylbenzyl ammonium chloride containing entrapped water pools. The reaction investigated was the back electron transfer from thionine radicals to aniline cation radicals formed via reductive quenching of triplet-state thionine by aniline. The recombination was shown to compete with the escape of the thionine radical from the water pool into the bulk solvent, i.e., benzene. Increasing the size of the water pool from 13 to 50 Å decreased the first-order rate constant for the back reaction from 10^8 to about 2×10^6 sec.$^{-1}$ Thus, the rate constant decreases with the third power of the reciprocal radius of the water pool. Earlier Monte Carlo calculations of Gösele et al.[78] predict such behavior. The results of these simulations indicate that the first-order rate constant of a diffusion-controlled reaction involving a radical pair is to a good approximation given by

$$\kappa = 4\pi \, DR/V_M \qquad (66)$$

where D and R are the sum of the diffusion coefficients and radii of the reactants, respectively, and V_M is the volume of the micelle available for diffusion.

The intramicellar back electron transfer was found to be impaired by applying a magnetic field. The effect of the magnetic field is to split the degeneracy of the triplet-state radical pair, reducing hyperfine interactions and hence intersystem crossing to the singlet system. Recombination is spin-allowed for the singlet but spin-forbidden for the triplet system. Hence, reduced intersystem crossing leads to a retardation in the recombination dynamics. Magnetic field effects have also been observed for charge recombination processes in normal micelles.[79] For a detailed discussion of these findings, the reader is referred to recent reviews.[11,80]

Multiphase systems such as inverted micelles belong to the class of water-in-oil microemulsions. Water is the dispersed microphase, while the hydrocarbon forms the bulk solvent. There is currently also a great deal of interest in oil-in-water microemulsions where an oil droplet of minute size is dispersed in water as the bulk phase.[81] These systems are rendered stable by the addition of a surfactant and cosurfactant. The main advantage of microemulsions over simple aqueous micelles is the high capacity for solubilization of hydrophobic compounds. A very large number of solubilizate molecules can be incorporated in the interior of the droplet without destabilizing the system. The surface potential of the droplets of ionic microemulsion, on the other hand, is decreased significantly with respect to that of micelles due to the presence of the cosurfactant, usually an alcohol, at the interface. As a consequence, the effects of the interfacial potential in assisting the charge separation are less pronounced in microemulsions. The low surface potential (<50 mV) also facilitates collision of droplets, enhancing the rate of exchange of solubilizate molecules between different droplets. Thus, in comparison to micelles, compartmentation of the reactants is less noticeable even in fast reactions, and the conventional kinetic laws for bulk systems are often applicable.

Jones and co-workers[82,84] were the first to make use of microemulsions for photochemical reactions using sensitizers such as Chl a or methyl red and crystal violet. In many of these studies a microemulsion consisting of sodium cetyl sulfate, pentanol, and hexadecane as surfactant, cosurfactant, and oil phase, respectively, was employed — which Jones et al. had developed and thoroughly characterized before. More recently, Dixit and Mackay[85] and Mackay and Grätzel[86] applied microemulsions with success to improve the efficiency of photogalvanic cells based on the iron-thionine reaction.

The investigations of Atik and Thomas[87] and Pileni and Chevalier,[88] as well as our own work,[89] have centered on the analysis of photoredox reactions and charge separation effects in microemulsion media.[87] An observation which is unique to this type of molecular assembly is that cation radicals formed from electron donors such as diphenylamine, dimethylnaphthalene (DMN), and TTF undergo dimer formation. For DMN the dimerization constant

$$DMN + DMN^+ \rightarrow (DMN)_2^+ \tag{67}$$

$K = 476$ M^{-1} has been evaluated.[90] The ease of dimer formation is likely to arise from the high local concentration of the substrate in the microemulsion droplet.

V. PHOTOINDUCED ELECTRON TRANSFER AND TRANSMEMBRANE REDOX REACTIONS IN VESICLES

A. Structural Features of Vesicles and Bilayer Lipid Membranes

Lipid vesicles, also known as liposomes, are closed bilayer membranes separating an occluded inner aqueous phase from the bulk solution (Figure 19). As in the case of micellar assemblies, they are constituted by amphiphilic compounds containing a hydrophilic and hydrophobic moiety. The preference of an amphiphile to aggregate into a liposome rather than a micelle is in general linked to the presence of two long alkyl chains in the molecule. For example, a very important class of natural liposome-forming agents consists of the phospholipids or phosphoglycerides which are found in cellular membranes. Among the

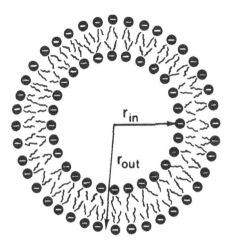

FIGURE 19.   Schematic illustration of the structure
of a bilayer vesicle.

phosphoglycerides present in higher plants and animals, phosphatidyl choline (lecithin)
derivatives such as 1-palmitoyl-2-oleoyl-phosphatidyl choline (PC)

$$H_3C-(CH_2)_{14}-\overset{\overset{O}{\|}}{C}-O-CH_2$$

$$H_3C-(CH_2)_7-\underset{H}{C}=\underset{H}{C}-(CH_2)_7-\overset{\overset{O}{\|}}{C}-O-\overset{|}{C}-H$$

$$H_2\overset{|}{C}-O-\overset{\overset{O}{\|}}{\underset{\underset{O^-}{|}}{P}}-O-CH_2-CH_2-\overset{+}{N}-\overset{CH_3}{\underset{CH_3}{CH_3}}$$

are most abundant. A space-filling model of phosphatidyl choline is shown in Figure 20.
The overall shape is roughly rectangular, the two fatty acid chains being approximately
parallel to each other, whereas the phosphoryl choline moiety points in the opposite direction.
   During the past years a large synthetic effort has been undertaken to provide artificial
analogues to these natural liposome-forming agents.[91,92] Prominent examples are the cationic
surfactant dioctadecyl dimethyl ammonium bromide (DODAC):

$$\begin{matrix} CH_3(CH_2)_{17} \\ CH_3(CH_2)_{17} \end{matrix} \overset{+}{\underset{N}{>}} \begin{matrix} CH_3 \\ CH_3 \end{matrix}$$

and anionic dihexadecyl phosphate (DHP):

$$\begin{matrix} CH_3-(CH_2)_{15}-O \\ CH_3-(CH_2)_{15}-O \end{matrix} P \overset{\overset{O}{\diagup}}{\underset{O^-}{\diagdown}}$$

   The formation of bilayer vesicles from these surfactants is a spontaneous process in aqueous
solution. As in the case of micelles, hydrophobic interactions are the major driving force
for aggregation. Water molecules are released from the hydrocarbon tails of the amphiphiles
as these tails become sequestered in the nonpolar interior of the bilayer, resulting in a large
entropy increase. Moreover, there are attractive van der Waals forces between the hydrocarbon tails which are closely packed in the membrane. Finally, there are favorable electrostatic and hydrogen-bonding interactions between the polar head groups and the surrounding

water molecules. All together, these factors afford a large gain in free energy which for vesicles is much higher than in the case of micelle formation. As a result, liposomes are formed at concentrations which are several orders of magnitude below the CMC of the single-chain surfactant analogues. Practically all the amphiphiles are in the aggregated state, the concentration of monomers being negligibly small.

In practice, liposomes are produced by suspending a suitable amphiphile, such as phosphatidyl choline, in an aqueous medium which thereafter is subjected to sonication to yield a dispersion of vesicles that are quite uniform in size. Alternatively, vesicles can be prepared by rapidly mixing a solution of lipids in ethanol with water. This can be accomplished by injecting the lipid through a needle. Vesicles formed by these methods are nearly spherical and have a diameter of 300 to 1000 Å, depending on the lipid used and other conditions, such as ionic strength and temperature.

In the electron-transfer experiments discussed below the solution composition in the interior aqueous compartment of the liposomes is frequently different from that of the exterior bulk phase. Such a condition can be readily achieved by forming the vesicles in the presence of the agent which is to be entrapped in the inside of the vesicles. The suspension is subsequently dialyzed or passed over a gel filtration column to remove the agent from the outside phase to which other substances can subsequently be added. An important question is how long such an unsymmetric distribution of solutes can be sustained across the membrane. The rate of migration through a bilayer membrane of the phosphatidyl choline type depends greatly on the nature of the diffusing molecule. Thus, the permeability coefficient for sodium ions is 10^{-12} cm/sec, while that for glucose is between 10^{-6} and 10^{-7} cm/sec. Surprisingly, the permeability of bilayer membranes for water is in general very high, the coefficient being 10^{-2} to 10^{-3} cm/sec. Since the width of the membrane is of the order of 40 Å, this implies that the transit time of a water molecule is at most a few hundred microseconds, while that of a sodium ion is about one hundred hours. Therefore, one can neglect on the time scale of the experiments dealing with photodriven electron transfer and in the absence of ionophores complications due to transmembrane diffusion of ionic species. It should be noted, however, that the permeability of bilayer vesicles with regard to certain ions may be dramatically increased under illumination with light in the presence of a membrane-bound sensitizer.[93]

To conclude this introductory section, it is worthwhile to point out that, apart from vesicles, photodriven electron transfer and transmembrane redox processes have also been studied with another type of artificial membrane, planar bilayers, which due to the color are sometimes referred to as black bilayer lipid membranes (BLMs). This type of structure can be formed across a small (1-mm) hole in a partition between two aqueous compartments (Figure 20). A fine paintbrush is dipped into a membrane-forming solution, such as phosphatidyl choline in decane. The tip is passed over the pinhole. Evaporation of the solvent leaves a bilayer membrane behind.

B. Photoinduced Electron-Transfer Reactions in Vesicle Dispersions

One of the first studies of light-induced redox reactions in vesicle systems was carried out by Chapman and Fast,[94] who reported cytochrome-c reduction (followed spectrophotometrically) in Chl-containing liposomes. A few years later, Tien et al.,[95-97] Ilani and Berns,[98] and Fong and Mauzerall[99] performed pioneering studies with pigments incorporated into BLMs. The key observation was the appearance of photocurrents when BLM, containing chloroplasts and separating two aqueous solutions of different redox potential, were illuminated by light. Similar experiments were soon performed with vesicles.[100-102] Figure 21 is a schematic illustration of the electron-transfer assay employed in these and many subsequent studies. The vesicle membrane, typically made up from egg lecithin or PC, contains a sensitizer, such as Chl. Electron donors and acceptors are added to different sides of the

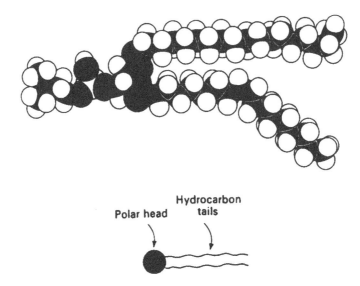

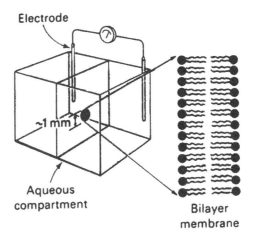

FIGURE 20.   Space-filing model of phosphatidyl choline molecules in the isolated state.

FIGURE 21.   Experimental setup for the study of planar bilayer membranes. The BLM is formed across a hole in a septum that separates two aqueous compartments. Heterogeneous photodriven electron transfer is studied by incorporating a sensitizer S into the BLM and adding suitable electron donors or acceptors to the aqueous phase. After flash excitation of the sensitizer the transient photocurrent (closed electrical circuit) or open-circuit photovoltage is recorded, providing a measure of the rate and efficiency of the interfacial charge transfer.

membrane. Frequently, a chemical potential gradient is set up across the membrane by dissolving one of the redox systems in the inner water core and the other in the outer aqueous phase. If the same redox couple is employed, the desired difference in chemical potential is obtained by adjusting appropriate concentrations of the oxidized and reduced form in the two compartments. Apart from the chemical potential, an electrostatic potential gradient can also be established across the membrane by using ionophores, such as valinomycin, which

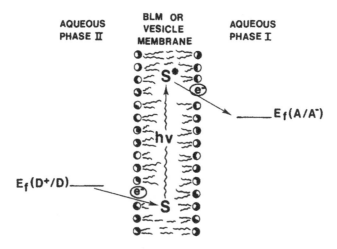

FIGURE 22. Typical electron-transfer assay in the light-driven transmembrane electron transfer in vesicle suspensions.

render the bilayer permeable with respect to potassium ions. The electrostatic potential
gradient is established as a consequence of the selective K^+ ion flux. For example, if
valinomycin-containing vesicles initially have K^+ dissolved in the inner water core, the
transport of K^+ to the outside produces excess negative charge in the inside, resulting in a
positive electrostatic potential gradient across the membrane. In the following, we take these
variations in the chemical and electrostatic potential into account by employing the Fermi
level or electrochemical potential of the redox couples in the thermodynamic description of
transmembrane electron-transfer reactions.

In Figure 22 the case is depicted where the electrochemical potential of the electron on
the acceptor side (II) is higher than that of the donor solution (I), E_f (redox II) $>$ E_f (redox
I). Therefore, light energy is required to drive the transmembrane redox process. The photons
are absorbed by the sensitizer, producing the reactive excited state S*, which in most cases
is the vibrationally relaxed singlet or triplet state. Note that in the excited state the sensitizer
is both a stronger reductant and oxidant as compared to the ground state. The respective
standard potentials can be estimated from the relations

$$E°(S*/S^+) = E°(S/S^+) - E(S*/S)$$ (68)

$$E°(S*/S^-) = E°(S/S^-) + E(S*/S)$$ (69)

where $E(S*/S)$ is the energy difference between the ground and reactive excited state. A
consequence of this shift in the redox potentials is that the excited sensitizer can undergo
either oxidative quenching by the acceptor:

$$(S*)_M + (A)_I \rightarrow (S^+)_M + (A^-)_I$$ (70)

or reductive quenching by the donor:

$$(S*)_M + (D)_{II} \rightarrow (S^-)_M + (D^+)_{II}$$ (71)

or both. Here, the subscripts M, I, and II indicate that the species is located in the membrane
or in one of the two aqueous phases, respectively. These photoevents are usually followed
by dark electron-transfer reactions regenerating the sensitizer:

$$(S^+)_M + (D)_{II} \rightarrow (S)_M + (D^+)_{II} \tag{72}$$

or

$$(S^-)_M + (A)_I \rightarrow (S)_M + (A^-)_I \tag{73}$$

The overall reaction corresponds in such a case to the light-driven transmembrane redox (TMR) process:

$$(A)_I + (D)_{II} \rightarrow (A^-)_I + (D^+)_{II} \tag{74}$$

In Table 1 we have listed experimental results obtained from the study of light-induced TMR in vesicle suspensions. From these investigations, the following features emerge:

1. Net oxidation-reduction occurs across liposomal and synthetic bilayer membranes and is stimulated by light if a suitable electron-transfer assay is employed in conjunction with a sensitizer.
2. In general, the quantum yields achieved for TMR are small. An exception is the PC-Chl a-glutathione system employed by Ford and Tollin,[114] which gave an overall efficiency of 20%, and the PC-ZnTPP-MV^{2+} system of Parmon et al.,[121] which gave 57%.
3. The quantum yields can be improved by solubilizing suitable electron relays, such as quinones or alloxazine derivatives, into the vesicle. The role of the relay is to carry the electron from the donor to the acceptor side of the membrane.
4. Electron exchange across the membrane has been suggested as the prominent mechanism for TMR in the absence of electron carriers. Figure 23 gives a schematic illustration of this process. Oxidative quenching forms the sensitizer cation radical on the side of the vesicle wall where the acceptor is located. Through the exchange reaction

$$S^+ + S \xrightarrow{k_{ex}} S + S^+ \tag{75}$$

the positive charge is transported to the opposite side, where it is used to oxidize the electron donor. The rate constants for the exchange reaction obtained from steady-state illumination and pulsed laser studies is about 10^4 sec^{-1} for PC vesicles and sensitizers such as Chl a and Ru(bipy)$_3^{2+}$. When this value is compared to the predictions of Equation 50 (in Chapter 1), using $\beta = 1.2$ Å^{-1} and $\Delta G^* = 4.8$ kcal/mol, which is the experimentally determined free energy of activation for the Ru(bipy)$_3^{3+}$-Ru(bipy)$_3^{2+}$ exchange reaction in water, one derives an electron-transfer distance of 10 Å. This is much shorter than the width of the membrane which is 40 to 50 Å and hence cannot be reconciled with a structural model where the sensitizer is located at the surface of the membrane. If the electron had to tunnel over a distance of 40 Å, the exchange rate should have been about a factor of 10^{13} times lower than the measured values. Hurst and Thompson[122] discussed this problem in a recent review and suggested that transmembrane electron transfer could conceivably occur in several steps by a "hopping" mechanism involving relatively deeply buried redox sites, thus decreasing the effective tunneling distance. Such configurations are likely to prevail in the case of the sensitizers employed in Table 1, i.e., Chl a and the long-chain derivative of Ru(bipy)$_3^{2+}$, which, due to the amphiphilic properties, are located in the palisade layer of the vesicle. A similar hopping mechanism appears to be operative in electron-transfer processes in monolayer assemblies (see Section VI).

Table 1
EXAMPLES OF ELECTRON-TRANSFER REACTIONS INDUCED BY LIGHT IN VESICLE SUSPENSIONS

Liposomes	Sensitizer	Acceptor (E°, NHE)	Donor (E°, NHE)	Results	φTMR[a]	Ref.
Egg lecithin	Chl a	Fe^{3+} acetate (0.34 V)	Ascorbic acid (−0.2 V)	Carotene required to observe TMR	0.075	73
PC		Fe^{3+} pyrophosphate (−0.14 V) MV^{2+} (−0.4 V) [b] AQS (−0.25 V)[c] NQ-OH (−0.15 V)[d] UQ 30 (0.1 V)[e] TMPD (0.27 V)[f]	No donor added; EPR detection of redox products	$Fe(CN)_6^{3-}$ not reduced by excited Chl a		
PC		$Fe(CN)_6^{3-}$ (0.36 V)	H_2O (0.82 V)	Proton transport agent added		76
		Cu^{2+}	Ascorbic acid (−0.2 V)	Carotene not required to observe TMR; $k_{et} = 6 \times 10^7$ mol⁻¹ sec⁻¹	2×10^{-6}	77
PC	Chl a	$Fe(CN)_6^{3-}$ (0.36 V)	H_2O	Proton carriers enhance TMR		78
Egg lecithin	Fast red A	AQS (−0.25 V)	Ascorbic acid (−0.2 V)	Surfactant derivative of AQS employed		79
PC	$Ru(bipy)_3^{2+}$[g]	$(C_7)_2V^{2+}$[g](−0.4 V)	EDTA	Electron exchange mechanism for TRM suggested; $k_{et} = 3 \times 10^4$ sec⁻¹	3.8×10^{-4}	80
	$Ru(bipy)_3^{2+}$[g]	$(C_7)_2V^{2+}$ (−0.4 V)	EDTA	Ionophores and transmembrane potential gradient used to enhance TMR	4.4×10^{-3}	81
	$Ru(bipy)_3^{2+}$	$(C_{12})_2V^{2+}$(−0.4 V)	EDTA triethanol amine (TEA)	Two-photon-activated transport across vesicle wall suggested		82
PC	$ZnC_{12}TPyP$[i]	AQS(−0.25 V)	EDTA	1,3-Dibutylalloxazine and 1,3-didodecylalloxazine increase yield of TMR; two-photon-activated transport postulated		83
Lecithin	Pyrene phenothiazine	H_2O	H_2O	Hydrated electrons observed as a result of photoionization		84
DODAC[j]	$Ru(bipy)_3^{2+}$[g]		MPTH[k]	Intravesicular electron transfer with subsequent ejection of MPTH⁺ in the aqueous phase		85
PC	Chl a	MV^{2+} (−0.4 V)	EDTA	Electron exchange model for TMR proposed; rate constant 10⁴ sec⁻¹	10^{-3} to 10^{-4}	86

Table 1 (continued)
EXAMPLES OF ELECTRON-TRANSFER REACTIONS INDUCED BY LIGHT IN VESICLE SUSPENSIONS

Liposomes	Sensitizer	Acceptor (E°, NHE)	Donor (E°, NHE)	Results	ϕTMR[a]	Ref.
	Chl a	5-(2-Methyl-1,4 naphthoqui-nonyl-3)-glutathione	Glutathione	TMR via electron exchange; k_{et} = 10^4 sec^{-1}	0.20	87
	Chl a	MV^{2+}	—	Inside vs. outside asymmetry in TMR observed	~0.1[l] ~0.01[m]	88
	Chl a	Benzoquinone	—	Charge separation affected by addition of anionic and cationic surfactants		89
	Chl a	Benzoquinone	—	Effect of salt addition to inner and outer water phase of redox products		90
	Chl a	MV^{2+} (−0.4 v) AQS (−0.25 V) Benzoquinone	—	Cholesterol increases forward and backward electron transfer rate		91
DHP[o]	Ru(bipy)$_3^{2+}$ ZnTMPyP^{4+n}	Methylene blue Fe(CN)$_6^{3-}$	Oxalate EDTA	Electron-transfer mediator added to membrane	0.1	92
	ZnTMP^{4+} ZnTPPS^{4-p}	Alkyl viologens Fe (CN)$_6^{3-}$	Amines	No transmembrane tunneling of electrons observed		93
PC	ZnTPP	MV^{2+} (−0.4 V)	EDTA	Two-photon-activated TMR	0.24 0.57^q	94
	Chl a	MV^{2+} (−0.4 V)	Ascorbate NADH	With ascorbate reverse TMR occurs in the dark		126

a   Quantum yield for transmembrane redox.
b   Methylviologen.
c   Anthraquinonedisulfonate.
d   Hydroxynaphthoquinone.
e   Ubiquinone.
f   N-Tetramethylparaphenyldiamine.
g   Amphipatic derivative of Ru(bipy)$_3^{2+}$.
h   Bis n-heptylviologen.
i   Amphipatic derivative of meso-tetra(4-pyridyl)porphyrin.
j   Dodecyl dimethyl ammonium chloride.

k   N-Methyl phenothiazine.
l   Quantum yield for electron transfer on outer vesicle surface.
m   Quantum yield for electron transfer on inner vesicle surface.
n   5,10,15,20-Tetrakis (N-methylpyridinium-4-yl)porphinatozinc.
o   Dihexadecyl phosphate.
p   5,10,15,20-Tetrakis(4-sulfonatophenyl)porphinatozinc.
q   In the presence of Ru(bipy)$_3^{2+}$.

a) ELECTRON EXCHANGE

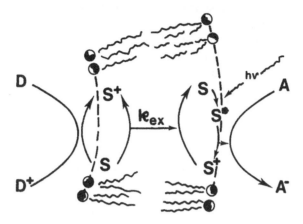

b) TWO PHOTON ACTIVATED

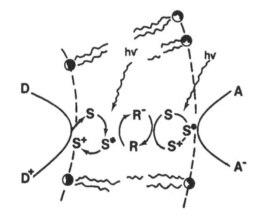

FIGURE 23. Schematic illustration of a two-photon-activated electron transfer across a vesicle membrane.

A number of very recent observations confirm the notion that an electron transfer across phospholipid membranes only occurs at a significant rate if the distance between donor and acceptor is less than 10 Å. At larger distances, the rate of TMR becomes very low. For example, Lee and Hurst[123] examined the electron exchange between 4-alkylpyridine pentaamine ruthenium ions located on opposite sides of the vesicle wall and found an exchange rate constant of 10^{-3} sec, which is much smaller than the values listed in Table 1. Tsuchida et al.[124] studied the distance dependence of electron transfer from liposome embedded alkanephosphocholine porphyrines to methylviologen. They found that the reactants are required to be within 12 Å in order for TMR to take place. Similar results have been obtained by Dannhauser et al.,[125] who concluded from studies of transmembrane electron exchange with polymer-linked manganese porphyrines that significant rates of charge transfer are observed only when the edge-to-edge distance separating two porphyrines is approximately 4 Å or less.

In the interpretation of TMR observed with vesicle suspensions, the ambiguity arises whether, instead of electron exchange, the charge transfer across the membrane occurs via diffusion. Ford et al.[107] have pointed out that the use of an amphipatic derivative of Ru(bipy)$_3^{2+}$ as a sensitizer rules out this possibility. With such a compound, the transmembrane diffusion should be orders of magnitude slower than the observed rate of electron exchange since a surfactant-type sensitizer would be required to undergo a "flip-flop" motion in order to migrate from one side of the membrane to the other. Hurst and Thompson[122] argue, on the other hand, that the binding of the amphiphilic sensitizer may perturb the membrane, both in the lateral plane and transversely, leading to enhanced membrane fluidity in the immediate binding environment. This would accelerate the rate of the flip-flop motion. Clearly, without direct and relevant information on the diffusional dynamics, the reaction mechanism for transmembrane electron transfer cannot be unambiguously assigned.

5.   A two-photon-activated transmembrane electron transfer has been observed in a few cases.[109-121] The reaction pathway in this case depends on whether the membrane contains, apart from the sensitizer, an electron carrier or not. In Figure 23 we outline the principles of carrier-mediated, two-photon-assisted TMR. In this case, the charge is displaced across the membrane by diffusion and not by electron transfer.

6.   The interfacial potential exerts an important effect on the quantum yield of light-induced charge separation. This is particularly evident from the flash photolysis studies of Ford and Tollin.

VI. ELECTRON TRANSFER IN MONOLAYLER ASSEMBLIES

It is inconceivable to omit a discussion of monolayer assemblies in a monograph dedicated to heterogeneous photochemical electron transfer. Although the technique for formation and manipulation of monolayer films has been known since the early studies of Blodget and Langmuir,[127] it is primarily the merit of Kuhn and collaborators[128] to have extended and utilized this technique to study energy and electron-transfer events in well-defined molecular arrays. Successive deposition of the same or different fatty acid molecules allowed the formation of assemblies with a well-defined structure and geometry. In Figure 24 we show the example of a Y-type multilayer structure composed of seven monolayers deposited in a head-to-head and tail-to-tail configuration. By incorporating amphiphiles with suitable redox function into such monolayer assemblies, it is possible to immobilize the reactants at known separation. The monolayer technique is therefore ideally suited for the examination of the distance effects in electron-transfer reactions.

These studies have shown that the electron-transfer rate depends in an exponential fashion on the distance separating electron donor and acceptor. This provides a direct confirmation of Equation 44 in Chapter 1, where the exponential decrease of the square of the matrix element with electron-transfer distance was postulated. Various values for the damping coefficient β have been determined. Mann and Kuhn[129] found $\beta = 1.5$ Å^{-1} from the change of the dark conductivity of one monolayer with the chain length of the fatty acid. Sugi et al.[130] derived $\beta = 1.2$ Å^{-1} from conductance measurement of multilayer arrays such as shown in Figure 24. The model usually applied to interpret charge transfer in such sandwich-type stacks of monolayers is adapted from the theory of dispersive transport of photocarriers in disordered solids. It is based on the notion that charge carrier trapping sites are located within the planes between adjacent layers. Typically, the density of those states is about 10^{15} eV^{-1} cm^{-2}. Electrons are transferred by tunneling from one interlayer plane to the next. The hopping rate across one layer is given by the Miller-Abrahams expression[131]

$$k_{et} = v_0(\beta R)^{3/2} \exp\left(-\beta R - \frac{\Delta E}{kT}\right) \qquad (76)$$

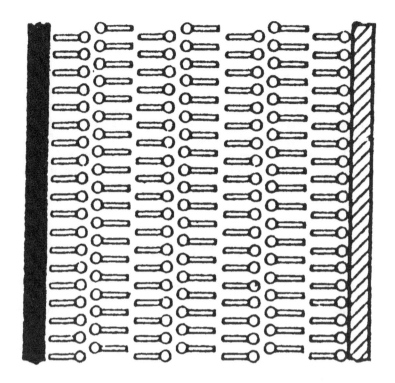

FIGURE 24. Schematic outline of a Y-type monolayer assembly consisting of seven monolayers sandwiched between two electrodes.

where ΔE is the activation energy for hopping and R is the tunneling distance. The trajectory the charge selects within the layer is obtained by minimizing the sum of ΔE and R, yielding:

$$k_{et} = v_0(\beta d)^{3/2} \exp(-\beta d - (2\beta/(\Pi^1 NdkT))^{1/2} \tag{77}$$

Here, d is the thickness of the monolayer and ^1N the density of interfacial states.

This model is also suited for explaining light-induced electron transfer within monolayer assemblies. For example, an amphiphilic cyanine dye acting as an electron donor in the photoexcited state has been incorporated into monolayer assemblies together with a viologen acceptor substituted by long alkyl chains.[132] The distance for electron transfer was varied by interspacing arachidic acid molecules between the layers containing the donor and acceptor, respectively. From the exponential decline of the electron-transfer rate with distance, a damping factor of 0.3 was obtained. Very recently, Nagamura et al.[133] derived a very small damping coefficient ($\beta = 0.005$) from transient photocurrent measurements in monolayer assemblies containing amphipathic porphyrins. These unusual observations were interpreted in terms of a superexchange mechanism.

VII. LIGHT-INDUCED CHARGE SEPARATION IN AQUEOUS DISPERSIONS OF INORGANIC COLLOIDS AND POLYELECTROLYTES

We terminate this chapter by briefly discussing the importance of polyelectrolytes and inorganic colloids in assisting light-induced charge separation. Polyelectrolytes are ionic water-soluble polymers. The charges produce electrolytic double layers much in the same way as in the case of micelles or vesicles. Similarly, inorganic colloids such as SiO_2 are charged in water, the surface potential depending strongly on the pH of the solution. The local field in the vicinity of the colloid can be exploited to achieve light-induced charge separation.[134-139] The elegant work of Loane et al.[136] and Willner et al.[137-138] using colloidal SiO_2 serves as an illustrative example.

In alkaline water, SiO_2 is negatively charged, the surface potential being -0.17 V. As a consequence, sensitizers such as $Ru(bipy)_3^{2+}$ or zinc meso-tetramethyl pyridiniumporphyrin ($ZnTMPyP^{4+}$) are strongly adsorbed at the surface of the particles. The acceptor employed, e.g., propylviologensulfonate, is neutral, becoming negatively charged upon photoinduced electron transfer. Thus, the reduced acceptor is repelled from the surface, impairing the back reaction. The improved charge separation in the presence of SiO_2 colloids manifests itself in a drastic augmentation of the quantum yield for the formation of reduced viologen.

REFERENCES

1. **Hately, M. D., Kozak, J. J., Rothenberger, G., Infelta, P. P., and Grätzel, M.,** *J. Phys. Chem.,* 84, 1508, 1980.
2. **Kalyanasundaram, K., Grieser, F., and Thomas, J. K.,** *Chem. Phys. Lett.,* 51, 101, 1974; **Humphrey-Baker, R., Moroi, Y., and Grätzel, M.,** *Chem. Phys. Lett.,* 58, 2, 1978.
3. **Rothenberger, G., Infelta, P. P., and Grätzel. M.,** *J. Phys. Chem.,* 83, 1871, 1979.
4. **Sandros, K. and Bäckström, H. L. J.,** *Acta Chem. Scand.,* 14, 48, 1960; *Acta Chem. Scand.,* 16, 938, 1962; **Sandros, K.,** *Acta Chem. Scand.,* 18, 2355, 1964.
5. **Yekta, A., Aikawa, M., and Turro, N. J.,** *Chem. Phys. Lett.,* 63, 543, 1979.
6. **Infelta, P. P. and Grätzel, M.,** *J. Chem. Phys.,* 70, 179, 1979.
7. **Infelta, P. P.,** *Chem. Phys. Lett.,* 61, 88, 1979.
8. **Kuanga, N., Selinger, B. K., and McDonald, R.,** *Aust. J. Chem.* 29, 1, 1976.
9. **Rodgers, M. A. J. and Wheeler, M. D. F.,** *Chem. Phys. Lett.,* 53, 165, 1978.
10. **Almgren, M., Grieser, F., and Thomas J. K.,** *J. Am. Chem. Soc.,* 101, 279, 1979.
11. **Turro, N. J., Grätzel, M., and Braun, A. M.,** *Angew. Chem. Int. Ed. Engl.,* 19, 6755, 1980.
12. **Grätzel, M. and Thomas, J. K.,** in *Modern Fluorescence Spectroscopy,* Wehri, E. L., Ed., Plenum Press, New York, 1976.
13. **Thomas, J. K.,** *Acc. Chem. Res.,* 10, 133, 1977.
14. **Thomas, J. K.,** *Chemistry of Excitation at Interfaces,* ACS Monograph Series, American Chemical Society, Washington, D.C., 1984.
15. **McQuarrie, D. A.,** *J. Chem. Phys.,* 38, 433, 1963.
16. **McQuarrie, D. A., Jachimooski, C. J., and Russel, M. E.,** *J. Chem. Phys.,* 40, 2914, 1964.
17. **McQuarrie, D. A.,** *Adv. Chem. Phys.,* 15, 149, 1969.
18. **Ishida, K.,** *J. Chem. Phys.,* 41, 2472, 1964.
19. **Darvey, I. G. and Ninham, B. W.,** *J. Chem. Phys.,* 45, 2145, 1966.
20. **Darvey, I. G. and Ninham, B. W.,** *J. Chem. Phys.,* 46, 1626, 1967.
21. **Staff, P. J.,** *J. Chem. Phys.,* 46, 2209, 1967.
22. **Thakur, A. K., Rescigno, A., and De Lisi, C.,** *J. Phys. Chem.,* 82, 552, 1978.
23. **Vass, Sz.,** *Chem. Phys. Lett.,* 70, 135, 1980.
24. **Rothenberger, G., Infelta, P. P., and Grätzel, M.,** *J. Phys. Chem.,* 85, 1850, 1981.
25. **Lachich, U., Ottolenghi, M., and Rabani, J.,** *J. Am. Chem. Soc.,* 99, 8062, 1977.
26. **Shinitzky, M.,** *Chem. Phys. Lett.,* 18, 247, 1973.
27. **Rothenberger, G.,** Fast Light Induced Reactions in Micellar Systems, thesis, EPFL, Lausanne, Switzerland, 1981.
28. **Koglin, P. K. F., Miller, D. J., Steinwandel, J., and Hauser, M.,** *J. Phys. Chem.,* 85, 2373, 1981.
29. **Matsuo, T., Aso, Y., and Kano, K.,** *Ber. Bunsenges. Phys. Chem.,* 84, 146, 1980.
30. **Ediger, M. D. and Fayer, M. D.,** *J. Chem. Phys.,* 78, 2518, 1983; **Ediger, M. D., Dominique, R. P., and Fayer, M. D.,** *J. Chem. Phys.,* 80, 1246, 1984.
31. **Hunter, R.,** *Foundations of Colloid Science,* Vol. 1, Oxford University Press, Oxford, 1986.
32. **Drummond, C. J., Grieser, F., and Healy, Th. W.,** *Faraday Discuss. Chem. Soc.,* 81, 0000, 1986.
33. **Bunton, C. A. and Moffatt, J. R.,** *J. Phys. Chem.,* 89, 4166, 1985.
34. **Pelizzetti, E. and Pramauro, E.,** in *Photoelectrochemistry, Photocatalysis and Photoreactors* (NATO ASI Ser. C.), Vol. 146, Schiavello, M., Ed., Reidel, Dordrecht, Netherlands, 1985, 271.
35. **Stigter, D.,** *J. Colloid Interface Sci.,* 32, 286, 1970.

36. Wallace, S. C., Grätzel, M., and Thomas, J. K., *Chem. Phys. Lett.*, 23, 359, 1973.
37. Grätzel, M. and Thomas. J. K., *J. Phys. Chem.*, 18, 2248, 1974.
38. Alkaitis, S. A. and Grätzel, M., *J. Am. Chem. Soc.*, 98, 3549, 1970.
39. Alkaitis, S. A., Grätzel, M., and Henglein, A., *Ber. Bunsenges. Phys. Chem.*, 79, 54, 1975.
40. Grand., D., Hautecloque, S., Bernas, A., and Petit, A., *J. Phys. Chem.*, 87, 5236, 1983.
41. Li, A. S. W. and Kevan, L., *J. Am. Chem. Soc.*, 105, 5752, 1983.
42. Ohta, N. and Kevan, L., *J. Phys. Chem.*, 89, 2415, 1985.
43. Chauvet, J. P., Viovy, R., Bazin, M., Santus, R., and Patterson, L. K., *Chem. Phys. Lett.*, 86, 135, 1982.
44. Bernas, A., Grand, D., Hautecloque, S., and Myasocdova, T., *Chem. Phys. Lett.*, 104, 105, 1984.
45. Narayana, P. A., Li, A. S. W., and Kevan, L., *J. Am. Chem. Soc.*, 104, 6502, 1982.
46. Plonka, A. and Kevan, L., *J. Phys. Chem.*, 89, 2087, 1985.
47. Humphrey-Baker, R., Braun, A. M., and Grätzel, M., *Helv. Chim. Acta*, 64, 2036, 1981.
48. Alkaitis, S. A., Beck, G., and Grätzel, M., *J. Am. Chem. Soc.*, 97, 5723, 1975.
49. Moroi, Y., Infelta, P. P., and Grätzel, M., *J. Chem. Phys.*, 69, 1522, 1978.
50. Razem, B., Wong, M., and Thomas, J. K., *J. Am. Chem. Soc.*, 100, 1629, 1978.
51. Waka, Y., Hamamoto, K., and Mataga, N., *Chem. Phys. Lett.*, 53, 242, 1978.
52. Grätzel, C. K. and Grätzel, M., *J. Phys. Chem.*, 86, 2710, 1982.
53. Wolff, C. and Grätzel, M., *Chem. Phys. Lett.*, 52, 542, 1977.
54. Moroi, Y., Braun, A. M., and Grätzel, M., *J. Am. Chem. Soc.*, 101, 5, 1979.
55. Moroi, Y., Infelta, P. P., and Grätzel, M., *J. Am. Chem. Soc.*, 101, 573, 1979.
56. Fendler, J. H., *Membrane Mimetic Chemistry*, John Wiley & Sons, New York, 1982.
57. Matsuo, T., *J. Photochem.*, 29, 41, 1985.
58. Humphry-Baker, R., Moroi, Y., Grätzel, M., Tundo, P., and Pelizzetti, E., *J. Am. Chem. Soc.*, 102, 3689, 1980.
59. Le Moigne, J., Gramein, P., and Simon, J., *J. Colloid Interface Sci.*, 60, 565, 1977.
60. Pileni, M. P., Braun, A. M., and Grätzel, M., *Photochem. Photobiol.*, 31, 423, 1980.
61. Brugger, P. A. and Grätzel, M., *J. Am. Chem. Soc.*, 102, 2461, 1980.
62. Brugger, P. A., Infelta, P. P., Braun, A. M., and Grätzel, M., *J. Am. Chem. Soc.*, 103, 320, 1981.
63. Infelta, P. P. and Brugger, P. A., *Chem. Phys. Lett.* 82, 462, 1981.
64. Shevalier, S., Lerebours, B., and Pileni, M. P., *J. Photochem.*, 27, 301, 1984; Lerebours, B., Chevalier, Y., and Pileni, M. P., *Chem. Phys. Lett.*, 117, 89, 1985.
65. Schmehl, R. H. and Whitten, D. G., *J. Phys. Chem.*, 85, 3473, 1981; Nagamura, T., Kurihara, T., Masuo, T., Sumitani, M., and Yoshinara, K., *J. Phys. Mem.*, 86, 4368, 1982.
66. Krasnovsky, A. A., in *Photosynthesis in Relation to Model Systems*, Barber, J., Ed., Elsevier/North-Holland, Amsterdam, 1979, chap. 9.
67. Kalyanasundaram, K. and Porter, P., *Proc. R. Soc. London Ser. A*, 364, 29, 1978.
68. Kalyanasundaram, K., Grätzel, M., and Pelizzetti, E., *Coord. Chem. Rev.*, 69, 57, 1986.
69. Luisi, P. L. and Straub, B. F., Eds., *Reverse Micelles*, Plenum Press, New York, 1984.
70. Menger, F. M., Donohue, J. A., and Williams, R. F., *J. Am. Chem. Soc.*, 95, 286, 1973.
71. Wong, M., Thomas, J. K., and Grätzel, M., *J. Am. Chem. Soc.*, 98, 2391, 1976.
72. Chance, B., *Proc. Natl. Acad. Sci. U.S.A.*, 67, 560, 1970.
73. Willner, I., Ford, W. E., Otrös, J. W., and Calvin, M., *Nature (London)*, 280, 830, 1979.
74. Mandel, D., Degani, V., and Willner, I., *J. Phys. Chem.*, 88, 4366, 1984.
75. Atik, S. S. and Thomas, J. K., *J. Am. Chem. Soc.*, 103, 3543, 1981.
76. Brochette, P., Zemb, T., Mathis, P., and Pileni, M. P., *J. Phys. Chem.*, 91, 1444, 1987; Brochette, P. and Pileni, M. P., *Nouv. J. Chim.*, 9, 551, 1985.
77. Ulrich, T. and Steiner, U. E., *Chem Phys. Lett.*, 112, 365, 1984.
78. Gösele, U., Klein, U. K. A., and Hauser, M., *Chem. Phys. Lett.*, 68, 29, 1979.
79. Tanimoto, Y., Shimidzu, K., Udagawa, H., and Hoh, M., *Chem. Lett.*, 153, 1983.
80. Kalynasundaram, K., Ed., *Photochemistry in Microheterogeneous Systems*, Academic Press, Orlando, Fla., 1987.
81. Shinoda, K. and Friberg, S., *Adv. Colloid Interface Sci.*, 4, 281, 1975.
82. Jones, C. E. and Mackay, R. A., *J. Phys. Chem.*, 82, 63, 1978.
83. Jones, C. E., Jones, C. A., and Mackay, R. A., *J. Phys. Chem.*, 83, 805, 1979.
84. Jones, C. A., Weaner, L. E., and Mackay, R. A., *J. Phys. Chem.*, 84, 1495, 1980.
85. Dixit, N. S. and Mackay, R. A., *J. Phys. Chem.*, 86, 4593, 1982.
86. Mackay, R. A. and Grätzel, M., *Ber. Bunsenges. Phys. Chem.*, 89, 526, 1985.
87. Atik, S. S. and Thomas, B. K., *J. Am. Chem. Soc.*, 103, 3550, 1981.
88. Pileni, M. P., *Chem. Phys. Lett.*, 75, 540, 1980; Pileni, M. P. and Chevalier, S., *J. Colloid Interface Sci.*, 92, 326, 1983.

89. Kiwi, J. and Grätzel, M., *J. Am. Chem. Soc.*, 100, 6314, 1978; Kiwi, J. and Grätzel, M., *J. Phys. Chem.* 84, 1503, 1980; Grätzel, C. K. and Grätzel, M., *J. Phys. Chem.*, 86, 2710, 1982.
90. Grätzel, C. K., Kira, A., Jirousek, M., and Grätzel, M., *J. Phys. Chem.*, 87, 3983, 1983.
91. Kunitake, T. and Okahata, Y., *J. Am. Chem. Soc.*, 97, 3860, 1977.
92. Fendler, J. H., *Acc. Chem. Res.*, 13, 7, 1980.
93. Lee, L. Y. C., Hurst, J. K., Politi, M., Kurihara, K., and Fendler, J. H., *J. Am. Chem. Soc.*, 105, 370, 1983.
94. Chapman, D. and Fast, P. G., *Science*, 160, 188, 1968.
95. Van, N. T. and Tien, H. T., *J. Phys . Chem.*, 74, 3559, 1970; Tien, H. T. and Varma, S. P., *Nature (London)*, 227, 1232, 1970.
96. Tien, H. T., *Bilayer Lipid Membranes: Theory and Pactice*, Marcel Dekker, New York, 1974.
97. Tien, H. T., *Bioelectrochem. Bioenerg.*, 9, 559, 1982.
98. Ilani, H. W. and Berns, D. S., *J. Membr. Biol.*, 8, 333, 1972.
99. Fong, F. and Mauzerall, D., *Nature (London) New Biol.*, 240, 154, 1972.
100. Mandel, M., *Biochim. Biophys. Acta*, 430, 459, 1976.
101. Oettmeyer, W., Norris, J. R., and Kate, J. J., *Z. Naturforsch.*, 31c, 163, 1976.
102. Tomkiewicz, M. and Corker, G. A., *Photochem. Photobiol.*, 22, 249, 1975.
103. Toyoshima, Y., Morino, M., Motoki, H., and Sukigara, M., *Nature (London)*, 265, 189, 1977.
104. Kurihara, K., Sukigara, M., and Toyoshima, Y., *Biochim. Biophys. Acta*, 547, 117, 1979.
105. Kurihara, K., Toyoshima, Y., and Sukigara, M., *Biochem. Biophys. Res. Commun.*, 88, 320, 1979.
106. Sudo, Y. and Toda, F., *Chem. Lett.*, 1011, 1978.
107. Ford, W. E., Otvos, J. W., and Calvin, M., *Proc. Natl. Acad. Sci. U.S.A.*, 76, 3590, 1980.
108. Laane, C., Ford, W. E., Otvos, J. W., and Calvin, M., *Proc. Natl. Acad. Sci. U.S.A.*, 78, 2017, 1981.
109. Matsuo, M., Takuma, K., Tsutsui, Y., and Nishigima, T., *J. Coord. Chem.*, 10, 187, 1980.
110. Matsuo, T., Itoh, K., Takuma, K., Hashimoto, K., and Nagamura, T., *Chem. Lett.*, 1009, 1980.
111. Barber, D. J. W., Norris, D. A. N., and Thomas, J. K., *Chem. Phys. Lett.*, 37, 481, 1976.
112. Infelta, P. P., Grätzel, M., and Fendler, J. H., *J. Am. Chem. Soc.*, 102, 1479, 1982.
113. Ford, W. E. and Tollin, G., *Photochem. Photobiol.*, 35, 809, 1982.
114. Ford, W. E. and Tollin, G., *Photochem. Photobiol.*, 38, 441, 1982.
115. Ford, W. E. and Tollin, G., *Photochem. Photobiol.*, 36, 647, 1982.
116. Fang, Y. and Tollin, G., *Photochem. Photobiol.*, 39, 429, 1983.
117. Fang, Y. and Tollin, G., *Photochem. Photobiol.*, 39, 685, 1984.
118. Ford, W. E. and Tollin, G., *Photochem. Photobiol.*, 40, 249, 1984.
119. Yablowska, E. E. and Shaflrovich, V. Ya., *Nouv. J. Chim.*, 8, 117, 1984.
120. Hurst, J. K., Lee, L. Y. C., and Grätzel, M., *J. Am. Chem. Soc.*, 105, 7048, 1983.
121. Parmon, V. N., Lymar, S. V., Tsvetkov, I. M., and Zamaraeo, K. I., *J. Mol. Cat.*, 21, 353, 1983.
122. Hurst, J. K. and Thompson, D. H. P., *J. Membr. Sci.*, 28, 3, 1986.
123. Lee, L. Y. C. and Hurst, J. K., *J. Am. Chem. Soc.*, 106, 7411, 1984.
124. Tsuchida, E., Kaneka, M., Nishide, H., and Itoshino, M., *J. Phys. Chem.*, 90, 2283, 1986.
125. Dannhauser, T. J., Nanog, M., Oku, N., Anzai, K., and Loach, P. A., *J. Am. Chem. Soc.*, 108, 5865, 1986.
126. Krasnovsky, A. A., Semenova, A. N., and Nikandrov, V. V., *Photobiochem. Photobiophys.*, 4, 227, 1982.
127. Blodget, K. B. and Langmuir, I., *Phys. Rev.*, 51, 964, 1935.
128. Kuhn, H., Böbins, D., and Bücher, H., in *Physical Methods of Chemistry*, Vol. 1 (Part 3), Weissberger, A. and Rossiter, B. W., Eds., Wiley-Interscience, New York, 1972, 577.
129. Mann, B. and Kuhn, H., *J. Appl. Phys.*, 42, 4348, 1971.
130. Sugi, M., Fuleni, T., Iizima, S., and Iriyama, K., *Bull. Electrotech. Lab.*, 43, 625, 1974.
131. Miller, A. and Abrahams, E., *Phys. Rev.*, 120, 745, 1960.
132. Kuhn, H., *J. Photochem.*, 10, 111, 1970.
133. Nagamura, T., Matano, K., and Ogawa, T., *J. Phys. Chem.*, 91, 2019, 1987.
134. Mesel, D. and Matheson, M., *J. Am. Chem. Soc.*, 99, 6577, 1977.
135. Sassoon, R. E. and Rabani, J., *J. Phys. Chem.*, 84, 1319, 1980.
136. Loane, C., Willner, I., Otros, J. W., and Calvin, M., *Proc. Natl. Acad. Sci. U.S.A.*, 78, 5928, 1981.
137. Willner, I., Otvös, J. W., and Calvin, M., *J. Am. Chem. Soc.*, 103, 3203, 1981.
138. Willner, I., Yang, J. M., Loane, C., Otvös, J. W., and Calvin, M., *J. Phys. Chem.*, 85, 3277, 1981.
139. Wheeler, J. and Thomas, J. K., *J. Phys. Chem.*, 86, 4540, 1980.

Chapter 3

CHARGE TRANSFER REACTIONS IN SEMICONDUCTOR SYSTEMS

I. INTRODUCTION

We shall now turn our attention to semiconductor systems and discuss the salient features of electron-transfer processes in this type of heterogeneous medium. Semiconductors are a very important class of solids and will be in the focus of our considerations. With regard to light-induced charge separation, they have several advantages over the molecular assemblies dealt with in the previous chapters. For example, the diffusion of mobile charge carriers in semiconductors is very fast. Even for a material such as TiO_2, which is characterized by a heavy effective electron mass, the diffusion constant of the electron is at least 10^4 times larger than that of a molecular charge carrier in a micelle or vesicle. This presents an important advantage for heterogeneous photoreactions, where the achievement of high efficiency requires rapid displacement of photogenerated species from the interior of the light harvesting unit to the surface. Furthermore, since the chemical transformations usually involve redox reactions at the surface, they can be mediated by deriving the semiconductor with suitable functional groups or by deposition of catalysts. Before addressing these issues in more detail, we shall first discuss the basic properties of semiconductor particles and give an overview of the experimental techniques that have been developed to investigate light-induced redox reactions in this type of assembly.

II. PROPERTIES OF SEMICONDUCTOR PARTICLES

The properties of semiconductor particles which are of interest here are listed in Table 1. They are divided into three categories: structural, optical, and electronic aspects. These will now be discussed in more detail.

A. Structural Properties
1. Size and Shape: Macroparticles, Colloids, and Q Particles
Particulate systems are commonly distinguished by size, and in Figure 1 we have applied this classification to semiconductors. The two major classes to be discerned are colloids and macroparticles. The latter have a size exceeding approximately 10^3 Å and form turbid suspensions. Colloids are smaller particles and give clear solutions. Among the colloids we distinguish semiconductors with normal optical and electronic behavior from those that display quantum size effects (Q particles). Quantum size effects occur when the Bohr radius of the first exciton in the semiconductor becomes commensurate with or larger than that of the particle. They are manifested in a blue shift of the fundamental absorption edge and of the luminescence maximum with decreasing particle size.[1] Since the Bohr radius of the charge carriers is related to the respective effective mass,

$$r_B = \frac{h^2 \epsilon_0 \epsilon}{c^2 \Pi^* m} \tag{1}$$

it depends on the semiconductor material. For example, in the case of CdS, $*m_{e-} = 0.2$ m_{e-}, $\epsilon = 8.9$, and $r_B = 24$ Å; while for TiO_2, $*m_{e-} = 30$ m_{e-}, 1-, and $\epsilon = 170$, yielding $r_B = 3$ Å. Thus, quantum size effects are expected for particles with radii below 25 and 3 Å for CdS and TiO_2, respectively, neglecting contributions from valence band holes.

As for the shape and crystalline structure of semiconductor particles, these are strongly

Table 1
PROPERTIES OF SEMICONDUCTOR PARTICLES

Structural	Optical	Electronic
Size	Absorption	Conduction
Shape	Luminescence	Intrinsic
Latticete type	Light-scattering (Raleigh, Mie)	n-Type
		p-Type
Lattice defects		Fermi level
		Charge carrier
		Mobility
		Band positions
		Charge carrier
		Recombination
		Doping

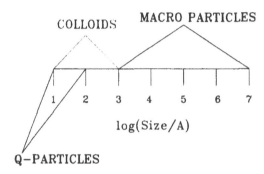

FIGURE 1.   Classification of semiconductor particulate systems.

influenced by the preparation methods. If so desired, materials with uniform size and high crystallinity can be prepared. Chemical precipitation or controlled hydrolysis and subsequent polymerization are the most frequently applied methods. Depending on preparation conditions, one obtains semiconductor particles of spherical shape with amorphous or crystalline structure. For example, we report in Figure 2 electron microscopy and quasi-elastic light scattering results obtained from a hydrosol of TiO_2. The latter was prepared by hydrolysis of $TiCl_4$ at a low temperature.[2] These data reveal that there is a distribution of particle sizes, the average diameter being 12 nm.

A particularly interesting development during recent years is the synthesis of monodisperse inorganic particles. Earlier work in this field concerned insulators, such as SiO_2[3] or noble metals, such as Pt or Au.[4] More recently, Matijevic[5] has elaborated methods for the preparation of monodispersed sols for a large number of compounds, including semiconducting oxides and chalcogenides. Apart from the application in the fabrication of special ceramics, such aggregates are also attractive from the viewpoint of heterogeneous catalysis and electron-transfer reactions. An important advantage is that these particles can be produced with extremely narrow size distribution and well-defined geometry. For example, it is possible to obtain spherically shaped aggregates with a smooth surface. The monodisperse nature of these sols facilitates the kinetic analysis of energy-and electron-transfer processes, whose rate is particle-size dependent. Moreover, applications in heterogeneous catalysis and light-energy conversion can also be envisaged. These particles could constitute suitable building blocks or supports for more complex units. The precise geometry should allow for the engineering of systems with suitable functionality, optimizing the performance in catalysis and heterogeneous electron transfer.

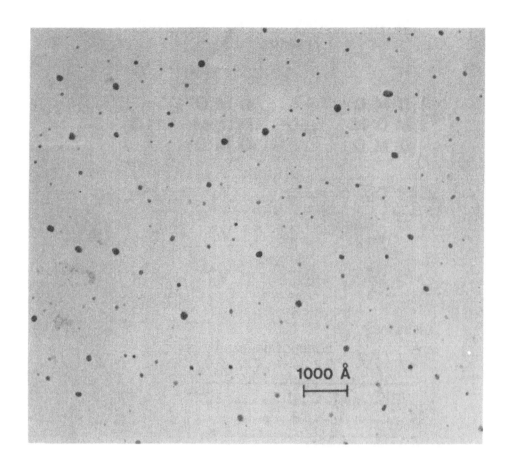

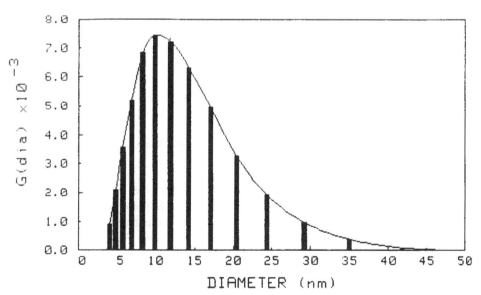

FIGURE 2. Size distribution and shape of colloidal TiO_2 particles. Top: electron microscopy and bottom: quasi-elastic light-scattering results, size distribution functions.

OXYGEN VACANCIES AS ELECTRON DONORS

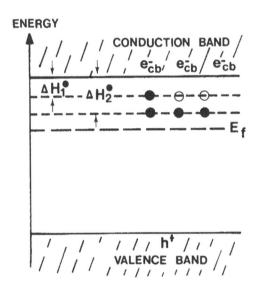

FIGURE 3.   Oxygen vacancies as electron donors.

2. Lattice Defects

While it is true that even very small semiconductor particles can be produced with a very high degree of crystallinity, occurrence of lattice defects is common and needs to be taken into consideration. In particular, there is a link between the nature of the defect and the character of the electronic conduction in the particles, which we shall briefly elaborate on in the following section. We shall restrict ourselves to the discussion of the most important point defects in oxides and chalcogenides, i.e., point defects such as anion and cation vacancies as well as interstitial cations. The formation of interstitial anions is energetically unfavorable and will not be considered. In order to establish a link between defect structure or deviation from stoichiometry and electronic properties, we examine the case of an oxide semiconductor of the general formula MO (Figure 3). The formation of an oxygen vacancy can be written in the Kroger notation as

$$O_o \rightleftharpoons V_o + \frac{1}{2}O_2 \qquad (2)$$

where O stands for an oxygen atom on the lattice site. The filled vacancy acts as an electron donor and ionizes in two subsequent steps:

$$V_0 \leftrightarrows V_0^{\cdot} + e_{cb}^{-} \tag{3}$$

$$V_0 \leftrightarrows V_0^{\cdot\cdot} + e_{cb}^{-} \tag{4}$$

Thus, an oxide with oxygen vacancies is an n-type conductor. Adding Equations 2 and 4 yields

$$O_O \leftrightarrows V_0^{\cdot\cdot} + 2e_{cb}^{-} + \tfrac{1}{2}O_2 \tag{5}$$

for which the mass action law gives

$$K = [V_0^{\cdot\cdot}][e_{cb}^{-}]^2 \, P_{O_2}^{1/2} \tag{6}$$

At sufficiently high temperature the first ionization is complete. Under these conditions $2[V_0^{\cdot\cdot}] = [e_{cb}^{-}]$, and therefore $[e_{cb}^{-}] = (2K)^{1/3}p^{-1/6}$. From this equation it is seen that the electron concentration, and hence the electrical conductance, of the powder should decrease with the 1/6 power of the partial pressure of oxygen, which is confirmed experimentally. Similarly to the oxygen vacancy, an interstitial metal is a donor center rendering the semiconductor n-conducting. Conversely, cation vacancies or acceptors render the oxide p-conducting. This is illustrated in Figure 4. Applying considerations analogous to those above for the metal vacancies, one finds that the concentration of holes, and hence the conductance of the semiconductor, should increase with the 1/6 power of the oxygen partial pressure. The relation between the deviation from stoichiometry, defect structure, and electronic properties are summarized in Table 2.

B. Optical Properties
1. Optical Absorption of Bulk Semiconductors and Colloidal Particles: Mie Theory

Semiconductors absorb light below a threshold wavelength, λ_g, the fundamental absorption edge, which is related to the band gap energy via

$$\lambda_g \text{ (nm)} = 1240/E_g \text{ (eV)} \tag{7}$$

Within the semiconductor, the extinction of light follows the exponential law

$$I = I_0 \exp(-\alpha l) \tag{8}$$

where l is the penetration distance of the light and α the reciprocal absorption length. For example, for TiO_2, α has the value 2.6×10^4 cm^{-1} at 320 nm, which implies that the light of wavelength 320 nm is extinguished to 90% after traversing a distance of 3900 Å.

Near threshold, the value of α increases with increasing photon energy. Frequently, a function of the type

$$\alpha \times h\nu = \text{const } (h\nu - E_g)^n \tag{9}$$

gives a satisfactory description of the absorption behavior in this wavelength domain. Here, the exponent has the value $\frac{1}{2}$ for a direct transition and 2 for an indirect one. A direct transition is characterized by the fact that, in the electronic energy vs. wave vector diagram,

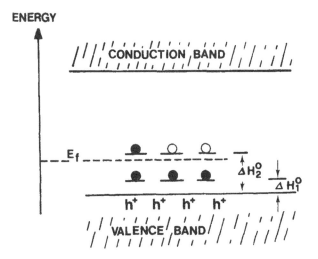

$$MM \rightleftharpoons V_M + M_{gas}$$

$$M_{gas} + \tfrac{1}{2} O_2 \longrightarrow M_M + O_o$$

$$\tfrac{1}{2} O_2 \rightleftharpoons V_M + O_o$$

$$V_M \rightleftharpoons V_M' + h^+ \qquad \Delta H_1^o$$

$$V_M' \rightleftharpoons V_M'' + h^+ \qquad \Delta H_2^o$$

FIGURE 4.   Cation vacancies as electron acceptors.

Table 2
DEVIATIONS FROM STOICHIOMETRY: OXIDES

Deficiency of O_2	Excess of O_2
TiO_{2-x}	NiO_{1+x}
ZnO_{1-x}	Ag_2O_{1+x}
Fe_2O_{3-x}	
Point defects	Point defects
Oxygen vacancies (V_O)	Metal vacancies (V_M)
Interstitial metal (M_i)	Acceptor centers
Donor centers	
n-Type conduction	p-Type conduction

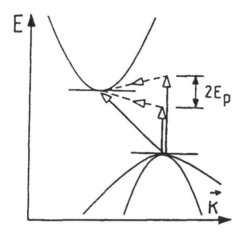

FIGURE 5. Energy vs. wave vector diagram illustrating the band structure for a semiconductor with an indirect gap. E_p is the contribution of lattice vibrations to the transition energy.

the minimum in the conduction band states is placed vertically above the maximum in the valence band energy states (Figure 13, Chapter 1). By contrast, for an indirect transition the two extrema are displaced from each other. As a consequence, the threshold excitation requires a contribution of lattice phonons in order to compensate for the change in the momentum vector during the transition (Figure 5). This reduces the absorption cross section and hence the value of α.

In some cases, deviations from Equation 9 have been observed. For example, Dutton[6] found that, for CdS, the reciprocal absorption length near band edge showed an exponential dependence on photon energy:

$$\ln \alpha = \beta h\nu/kT \qquad (10)$$

The coefficient β had a value of 2.1 for single crystals with hexagonal structure. In our laboratory,[7] colloidal particles of CdS have been produced by adding H_2S to an aqueous solution of $Cd(NO_3)_2$ (method 1) or by rapid mixing of NaS and $Cd(NO_3)_2$ solutions (method 2). It was discovered that hexametaphosphate was suited to stabilize the ultrafine particles, preventing the aggregation by electrostatic protection. The two methods yield particles with an average size of 50 and 30 nm, respectively. Each particle is a single crystal. Interestingly, the reciprocal absorption length near band edge also followed the Dutton law. However, the coefficient was found to have a significantly smaller value that observed for single crystals. This was attributed to the large fraction of CdS atoms which are present at the surface and hence exposed to the solvent in the colloidal solutions. Support for this interpretation comes from the observation that the value β decreases with the proportion of CdS that is present at the particle surface.[8]

In dilute solution colloidal semiconductor particles remove light from the incident beam by scattering and absorption. The whole extinction spectrum can be calculated from the Mie theory.[9] Mie obtained a rigorous solution of absorption and scattering by a single sphere. This can be applied to a collection of spheres if a number of conditions are satisfied.[10] Thus, the distance between spheres must be larger than the wavelength so that the spheres scatter independently, and the spheres must be randomly oriented. If, in addition, the particles are much smaller than the incident light ($R \ll \lambda$) one obtains for the reciprocal absorption length (centimeter) of the light in a solution containing the colloid

$$\alpha = \frac{18\Pi c_p V_p n_s^3 \epsilon_2}{\lambda(\epsilon_1 + n_s^2)^2 + \epsilon_2^2} \qquad (11)$$

where

$$\epsilon = \epsilon_1 + i\epsilon_2 \qquad (12)$$

is the complex dielectric constant of the particle, c_p is the particle concentration expressed in number of particles per cubic centimeter, V_p is the volume of one particle, n_s is the refractive index of the solvent in which the colloid is dispersed, and λ is the wavelength of the incident light.

The significance of the dielectric constants ϵ_1 and ϵ_2 becomes clearer by expressing ϵ in terms of the complex refractive index of the particle:

$$\epsilon = (n_p + ik)^2 = n_p^2 - k_p^2 + i(2k_p n_p) \qquad (13)$$

where n_p is the real refractive index of the particle and k_p is the absorption index, which is proportional to the reciprocal absorption length α_p of the light with wavelength λ within the particle:

$$k_p = \alpha_p \lambda/4\Pi \qquad (14)$$

A comparison of Equations 13 and 12 identifies ϵ_1 with $n_p^2 - k_p^2$ and ϵ_2 with $2n_p k_p$.

Mie's theory has been widely applied to interpret the extinction spectra of colloidal dispersions, in particular metal sols.[10] For example, the brilliant ruby colors of colloidal golds as well as the yellow color of colloidal silver are well explained by this electromagnetic theory. The advent of experimental techniques to prepare monodispersed sols has allowed one to study the effect of particle size on the extinction spectra. In this regard, a recent investigation by Hsu and Matijevic[11] on colloidal α-Fe$_2$O$_3$, a semiconductor with a band gap of 2.2 eV, is noteworthy. Within the 0.1- to 0.16-nm size range, the optical spectra of these hydrosols agreed well with the predictions of the Mie theory.

2. Optical Absorption of Very Small Particles: Quantum Size Effects

At very small particle radii, deviations from the Mie equation occur, and this is due to the size quantization effects which were alluded to already earlier on. Size quantization effects have been widely investigated in the past for metal particles. In bulk metals, considered as the limit of an infinitely large system, there is a continuous spectrum of electronic states. The distribution function of these states was discussed in Chapter 1. In contrast, in very small colloidal particles the energy levels are quantized, the spacing between adjacent states being of the order of E_f/N, where E_f is the Fermi energy and N is the number of atoms in one particle. Since E_f has a value of a few electronvolts, the energy levels in a particle containing, say, 10^4 atoms are separated by about 10^{-4} eV. This quantization of states produces a number of anomalies in the properties of the particles, including the optical absorption. This fact was already recognized some 50 years ago by Fröhlich.[12] The more recent literature is covered in a review by Kubo et al.[13]

Effects arising from the spatial confinement of charge carriers in semiconductors have also been the subject of intensive investigations. Much of the effect has concentrated on two-dimensional structures, such as quantum wells and superlattices.[14] In thin films of semiconductors a size quantization perpendicular to the film occurs. Earlier observations of confinement effects in three-dimensional systems concerned small CdS particles in glass

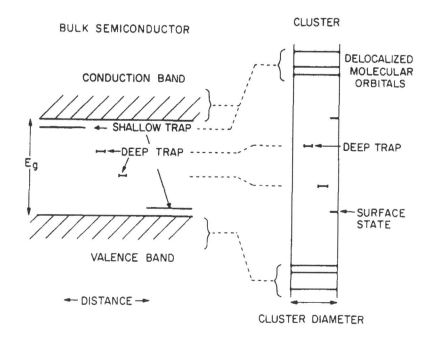

FIGURE 6.   Spatial electronic state correlation diagram for bulk semiconductors and clusters. (Adapted from Brus, L. E., *J. Chem. Phys.*, 80, 4403, 1984; *J. Chem. Phys.*, 79, 5566, 1983.)

matrixes[15] which show blue-shifted absorption and emission spectra. These effects were later interpreted in terms of the symmetry of the Cd^{2+} environment.[16] Systematic investigations of quantum size effects in small semiconductor particles date back to the work of Berry[17] on silver halide microcrystals. With decreasing size of the particles, a blue shift in the band edge absorption, as well as the disappearance of the exciton peak, was observed. This was correctly interpreted in terms of charge carrier confinement: below a certain size there is insufficient space for the electron-hole pairs formed during band gap excitation to coexist within the particles, leading to a change of optical properties. Similar observations were made by Stasenko,[18] Skomyakov et al.,[19] and Hayashi at al.[1] on CdS microcrystals.

In colloidal semiconductor particles the effect of local confinement of the charge carriers produces discrete electronic states in the valence and conduction bands and increases the effective band gap (Figure 6). To a first approximation, the energy of the quantized levels is inversely proportional to the effective mass and the square of the particle diameter. The consequences of size quantization for the absorption features of a semiconductor dispersion can be quite dramatic. Thus, HgSe colloids consisting of large particles (500 Å) are black, since the band gap of the bulk material is only 0.3 eV, corresponding to an absorption threshold at 4130 nm. Due to the very low effective electron mass, $*m_{e-} = 0.05\ m_{e-}$, the band gap of HgSe is strongly dependent on particle size. For example, the absorption threshold for 30-Å-sized particles is at 380 nm, implying that the band gap of the semiconductor has increased to 3.2 eV.[20] Similar effects have been observed with other materials, such as PbSe,[20] CdAs,[21] ZnO,[22] and ZnP.[23] An elegant method of controlling the size of the semiconductor particle during preparation is to effect the chemical precipitation in the water pockets of reversed micelles[24] or vesicles.[25]

Several attempts have been made to calculate the electronic energy levels in these Q particles. Brus[26] used an exciton model in order to derive the electronic energy states of an electron-hole pair confined to a small spherical semiconductor crystallite. A quantum mechanical calculation gave the lowest eigenstate of the Wannier exciton in such clusters,

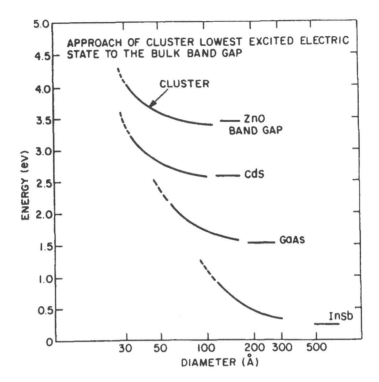

FIGURE 7.   Apparent band gap of a semiconductor cluster as a function of particle size. (Adapted from Brus, L. E., *J. Chem. Phys.*, 80, 4403, 1984; *J. Chem. Phys.*, 79, 5566, 1983.)

which corresponds to the apparent band gap (Figure 6). The exciton wave function was approximated by one or a few configurations of particle-in-a-box orbitals. The energy of the first excitonic state of a semiconductor cluster with radius R is given by the approximate expression

$$E_g(R) \simeq E_g(R = \infty) + \frac{h^2\pi^2}{2R^2}\left[\frac{1}{m_{e^-}} + \frac{1}{m_{h^+}}\right] - \frac{1.8\ e^2}{\epsilon R} \tag{15}$$

The first term in this equation is the band gap of the bulk semiconductor, the second term corresponds to the sum of the confinement energies for the electron and the hole, and the last is the coulombic interaction energy.

The coulomb term shifts E(R) to smaller energy as R, while the quantum localization terms shift E(R) to higher energy as R^2. As a result, the apparent band gap will always increase for small enough R, and this is shown in Figure 7 for various semiconductors.

Foitik et al.[27] have used an alternative model to calculate the size dependency of semiconductor colloids, where the confined exciton is treated as an electron with the reduced mass $\mu^{-1} = (1/m_{e^-} + 1/m_{h^+})$ which moves in the field of the walls, the hole being fixed in the center of the particle. Comparison with the experimental results obtained for CdS gives better agreement than Equation 15, which predicts too large shifts in the band gap with decreasing particle radius. However, as has been pointed out by Foitik et al.,[27] the agreement may be fortuitous since the errors introduced by the approximations (infinitely high potential energy wall at the particle surface, underestimation of the kinetic energy of the confined electron) cancel each other. Calculations with refined quantum mechanical models have recently been carried out.[28]

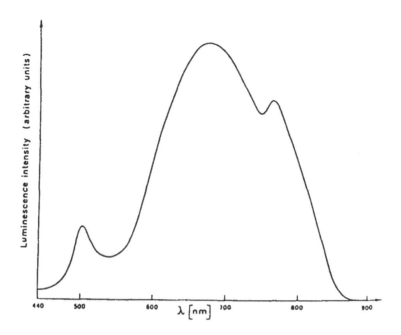

FIGURE 8. Typical luminescence spectrum of colloidal CdS particles in water. (From Ramsden, J. J. and Grätzel, M., *J. Chem. Soc. Faraday Trans. 1*, 80, 919, 1984.)

3. Photoluminescence of Semiconductor Particles and Electrodes

The luminescence of semiconductor dispersions has been investigated in great detail over the last few years following the initial work on aqueous CdS colloids.[29] While there is by now a wealth of experimental results available,[24,30] only a few authors[7,31,32] have attempted to assign the observed emissions. The earlier analysis[7] showed that the luminescence spectrum of the CdS particles contains frequently two bands: a green emission resulting from the recombination of free carriers and a red one attributed to sulfur vacancies. Typical results are presented in Figure 8.

Sulfur vacancies are common point defects in CdS, and the ready formation

$$CdS \rightleftarrows CdS_{1-x} + V_s \ (x \ll 1) \tag{16}$$

explains the n-type behavior of this semiconductor. In analogy to oxygen vacancies, V_s acts as a donor, and the standard enthalpies required for the two successive ionizations of this center are 0.03 and 0.7 eV, respectively.[33] Red luminescence arises from the reaction of photogenerated holes with singly ionized sulfur vacancies:

$$h^+ + V_s^+ \rightarrow V_s^{2+} + h\nu \ (\sim 700 \text{ nm}) \tag{17}$$

and hence is explained by the Lambe-Klick model (Figure 9). Both the green and the red emission are quite weak. However, the red emission can be enhanced drastically, for example, by the preparation of the colloid in a mixture of water and acetonitrile.[7]

A noteworthy feature of the luminescence of CdS and other II-VI or III-V semiconductors is that it is extremely sensitive to the presence of electron acceptors such as MV^{2+} or benzoquinone.[30] In water, a concentration of $10^{-8} M$ of MV^{2+} suffices to quench 50% of the red emission of CdS.[6] A quantitative analysis of the results, using Poisson statistics to describe the quencher distribution over the particles, showed that only one MV^{2+} molecule per aggregate is required to quench the emission.[6,7] The role of the MV^{2+} is to scavenge conduction band electrons, intercepting the radiative and nonradiative recombination with

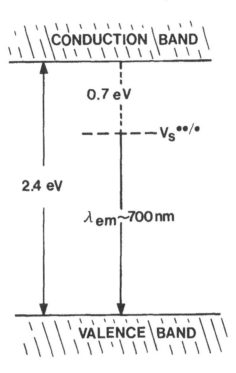

FIGURE 9.   Self-activated luminescence of CdS colloids, Lambe-Klick model for the operation of sulfur vacancies as luminescence center.

holes. Recent laser photolysis experiments with picosecond time resolution have shown that this reaction is very fast, requiring less than a nanosecond.[32,33]

Photoluminescence has also been extensively used to probe charge carrier recombination processes in semiconductor electrodes.[34,39] The scope of this technique is illustrated by the elegant work of Ellis et al.[37,39] on II-VI and III-V compounds. For example, CdS substitutionally doped with Te displays a photoluminescence at 600 nm. This arises from holes trapped at the Te centers which recombine radiatively with conduction band electrons via an excitonic mechanism. The emission intensity reflects the interplay of electron-hole separation and recombination in the semiconductor electrode which is in contact with an electrolyte. Thus, upon decreasing the band bending within the space charge layer of the semiconductor by changing the bias potential, the luminescence quantum yield increases strongly while the photocurrent decreases. The explanation of this effect is that with reduced band bending there is less driving force for electron-hole recombination and more likelihood for recombination. Furthermore, the emission intensity is sensitive to the penetration depth of the light. If the electrode is excited close to band edge, the light penetrates far into the semiconductor. Most of the charge carriers are generated beyond the space charge layer where the recombination probability is high, resulting in a relatively strong emission. By contrast, if the penetration depth of the light is commensurate with the depletion layer width, the electron hole pairs are separated by the local field, and very little radiative recombination occurs. Thus, photoluminescence is a useful complement to photocurrent measurements in that it provides important information on the dynamics of charge carrier reactions at the interface as well as the nature and width of the space charge layer.

Photoluminescence has also been applied to probe the nature of surface states present at the semiconductor electrolyte interface. Extensive studies have been performed by Nakato et al.[42-45] with materials such as TiO_2, ZnO, CdS, and GaP. For example, TiO_2 and GaP

ENERGY

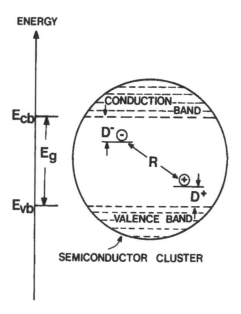

FIGURE 10.   Radiative and radiationless tunneling recombination of trapped electron-hole pairs. D^- and D^+ are the trap depth for the electron and hole, respectively.

show luminescence peaks at 1000 and 800 nm, respectively, which have been attributed to trapped hole states at the surface. In the presence of reducing agents such as $Fe(CN)_6^{4-}$ the luminescence was quenched, which corroborated this assignment.

4. Quantum Size Effects in the Photoluminescence of Colloidal Semiconductors

Due to the quantum effects discussed above, the luminescence of colloidal semiconductors, similar to the band edge absorption, is strongly influenced by the particle size. Typically, a broad emission is observed which blueshifts with a decreasing radius of the aggregate. Effects of this type were first analyzed by Papavassiliou[46] on extremely small CdS particles, and have been confirmed in the work of Brus[26], Foitik et al.,[21] Koch et al.,[22] Weller et al.,[23] and Nozik et al.,[18] and by our own studies.[7] The luminescence of these clusters is readily detectable at 298°K and can be quite strong at a cryogenic temperature.

Luminescence analysis provides an important tool for studying the dynamics of charge carrier recombination in such colloidal semiconductors. For CdS clusters in alcohol glasses, the emission decay time is wavelength and temperature dependent, but it becomes temperature independent below 30°K.[47] In this temperature range, the reciprocal luminescence lifetime, $1/\tau$, for 22-Å-sized clusters is 10^5 and 4.5×10^4 sec^{-1} at 580 and 400 nm, respectively. This observation can be rationalized in terms of radiative and radiationless recombination of trapped carriers (Figure 10). At low temperature, these two processes occur exclusively via a tunneling mechanism.

The contribution of the radiative channel to the overall decay can be estimated by using a model derived for bulk semiconductors.[46-51] The rate constant depends on the distance R separating the electron-hole pair via

$$k(R) \simeq 2\pi |H_{e^- - h^+}(R)|^2/\hbar \tag{18}$$

$$k(R) = 5 \times 10^8 \exp(-2R/a_o) \; sec^{-1} \tag{19}$$

These equations have the same form as Equation 43 in Chapter 1, the nuclear factor (FC) being set equal to 1. H is the exchange integral which measures the interaction energy between the donor and acceptor, i.e., the overlap between the tails of the wave function of the trapped electron and hole; a_0 is the radius of the hydrogenic radius of the wave function for the charge carrier residing in the trap with the lower binding energy.

The dominant route for charge carrier recombination in semiconductor clusters is non-radiative. This is due to the strong coupling of the wave functions for the trapped electrons and holes to lattice vibrations. The presence of strong vibronic interactions is manifested in the broad nature of the emission spectrum of semiconductor clusters. If there were no phonon interactions, a single emission line would be expected.

The rate constant for nonradiative exothermic tunneling is given by the general expression for nonadiabatic electron transfer: Equation 43 in Chapter 1. Using the fully quantum mechanical model discussed in Chapter 1, Section III.C in order to derive the Franck-Condon overlap integral, one obtains[52,53]

$$k_{nr} = \frac{2\pi}{\hbar} \mid H_{e^- - h^+} \mid^2 \exp[-S(2\bar{n} + 1)] \, I \, (2S[\bar{n}(\bar{n} + 1)]^{1/2} \times [^-(\bar{n} + 1)/\bar{n}]^{p/2} \quad (20)$$

which is the same as Equation 89 in Chapter 1.

Chestnoy et al.[47] have employed Equation 20 to evaluate the temperature dependence of the photoluminescence of CdS clusters with a size of 38Å (effective band gap, 3.2 eV) and 22 Å (effective band gap, 3.2 eV). For the larger particles the exothermicity of electron-hole recombination is 2.8 eV, while for the smaller particles it is 3.4 eV. Equation 20 provided a satisfactory description of the temperature effect on the luminescence lifetime of CdS clusters, supporting the notion that trapped electron-hole pairs recombine via a radiationless multiphoton process which is brought about by a strong coupling of the wave function of the trapped charge carriers to lattice vibrations. A best fit of the experimental results gave $h\nu = 0.017$ and 0.02 eV for the vibrational excitation energy of the reactant and product oscillators in the 38- and 22-Å-sized clusters, respectively. These values are within the known phonon spectrum of CdS. For the Franck-Condon displacement factor $S = \lambda/h\nu$ a value of 110 was derived corresponding to a reorganization energy of about 2 eV for the recombination process.

As for the wavelength dependence of the emission lifetime, this arises from the contribution of the coulombic energy of interaction between the electron-hole pair to the total energy of the emitted photon.

$$h\nu = E_g - (D_+ - D_-) + e^2/\epsilon R \quad (21)$$

where D_+ and D_- are the trap depths for the electron and hole, respectively, measured with respect to the lowest delocalized state.

It follows from Equation 20 that close pairs emit at higher energy than those which are further apart. Furthermore, for the close pairs the emission is faster since both k_r and k_{nr} increase with decreasing distance (Equations 18 and 20). Therefore, the emission lifetime becomes wavelength dependent, decreasing with increasing energy of the photons.

C. Electronic Properties
1. Band Edge Positions

Fundamental aspects of the electronic band structure in semiconductors were discussed in Chapter 1. In Figure 11, we list the band gap and band edge position for a number of ionic and covalent materials. The data refer to conditions where the semiconductor is in contact with an aqueous electrolyte of pH 1 and have been derived from flat band potential

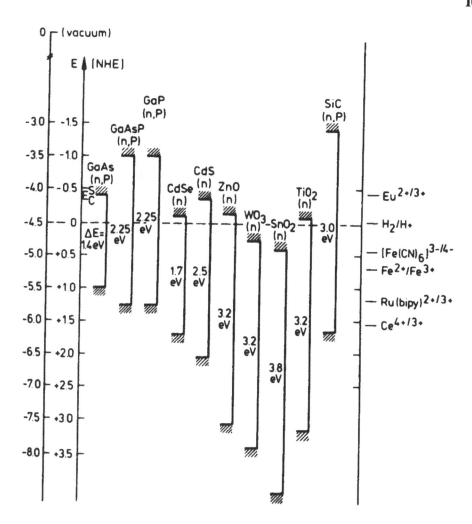

FIGURE 11. Band edge position of several semiconductors in contact with aqueous electrolyte at pH 1.

determinations via capacity measurements. The knowledge of the band edge position is particularly useful in photocatalysis. For example, in Figure 11 we have indicated the standard potentials for several redox couples which indicate the thermodynamic limitations for the photoreactions that can be carried out with the charge carriers. For example, if a reduction of the species in the electrolyte is to be performed, the conduction band position of the semiconductor must be positioned above the relevant redox level.

Note that the ordinate in Figure 11 presents internal and not free energy. The free energy of an electron-hole pair is smaller than the energy of the band gap. The reason for this behavior is that the electron-hole pairs have a significant configurational entropy arising from the large number of translational states accessible to the mobile carriers in the conduction and valence band.

The free energy of the charge carriers generated by photoexcitation of semiconductors is directly related to the chemical potential. It was shown in Chapter 1 that in the dark, under thermal equilibrium, the chemical potential of the electron is equal to that of the hole and corresponds to the Fermi level of the solid. The position of the Fermi level with respect to the band edges is given by Equation 129 in Chapter 1. Under illumination the system departs from equilibrium, and the chemical potential of the electron is different from that of the

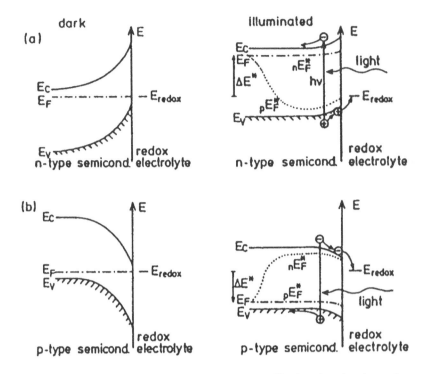

FIGURE 12.   Quasi-Fermi levels and photovoltages at an illuminated semiconductor-electrolyte contact with depletion layers. (a) n-type semiconductor electrolyte and (b) p-type semiconductor electrolyte.

hole. As a result, the Fermi level splits into two quasi-Fermi levels, one for the electron and one for the hole. This is illustrated schematically for an n-type semiconductor in Figure 12. The maximum work that can be done by an electron-hole pair is given by

$$w = \mu_{e^-} - \mu_{h^+} = \text{photovoltage} \times e \tag{22}$$

where e is the elementary charge. Using Equations 126 and 127 in Chapter 1 to express the chemical potential, one obtains

$$\mu_{e^-} = E_{cb} + kT \ln(c_{e^-}^* / {}^l n_{cb}) \tag{23}$$

and

$$\mu_{h^+} = E_{vb} - kT \ln(c_{h^+}^* / {}^l n_{vb}) \tag{24}$$

where $c_{e^-}^*$ and $c_{h^+}^*$ are the nonequilibrium concentrations of electrons and holes, respectively, which depend on light intensity. Since at low light level, where the optical transition is far away from saturation, the concentration of minority carriers is proportional to the light intensity I; w and the open circuit photovoltage increase with ln I.

2. Space Charge Layers and Band Bending

The generation of a space charge layer requires the transfer of mobile charge carriers between the semiconductor and the electrolyte. When an electroactive species is present in the electrolyte, the charge transfer can take place directly across the semiconductor solution interface. Alternatively, in the absence of a suitable redox couple in solution, the semicon-

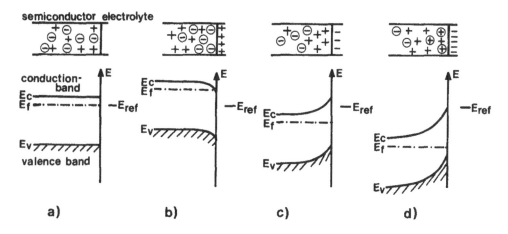

FIGURE 13. Space charge layer formation at the n-type semiconductor-solution interface. (a) Flat band situation; (b) accumulation layer; (c) depletion layer; and (d) inversion layer.

ductor can be polarized by applying an external bias voltage across the junction via an ohmic contact mounted at the back of the electrode. Within the space charge layer, the valence and conduction bands are bent. Here, four different situations may be envisaged. These are illustrated in Figure 13 for an n-type semiconductor in contact with an electrolyte. If there is no space charge layer, the electrode is at the flat band potential. If charges are accumulated at the semiconductor side which have the same sign as the majority charge carriers, one obtains an accumulation layer. If, on the other hand, majority charge carriers deplete into the solution, a depletion layer is formed. The excess space charge within this layer is given by immobile ionized donor states.

The depletion of majority carriers can go so far that the concentration at the surface decreases below the intrinsic level. If the electronic equilibrium is maintained, i.e., Equation 129 in Chapter 1 is obeyed, the concentration of holes in this region of the space charge layer exceeds that of electrons. As a consequence, the Fermi level is closer to the valence than the conduction band and the semiconductor is p-type at the surface and n-type in the bulk. This is called an inversion layer.

The illustration in Figure 13 refers to n-type semiconductors. For p-type materials, analogous considerations apply, holes being the mobile charge carriers and immobile, negatively charged acceptor states forming the excess space charge within the depletion layer. The bands bend downward in this case.

In order to derive an expression for the width of the depletion layer, we use the definition $dq = C_{sc}\, d\phi$, where C_{sc} is the capacity per unit area of the space charge layer, and q is the excess charge per unit area, and express it in the form:

$$q^2 = 2\epsilon\epsilon_o \int_{\phi_s}^{\phi_b} \rho(\phi)\, d\phi \qquad (25)$$

Here, ϕ_s and ϕ_b are the electrostatic potentials at the semiconductor surface and the bulk, respectively, and ρ is the charge density. Using the Poisson equation to express ρ and making the assumption that the density of ionized donors and acceptors is constant within the depletion layer, one obtains

$$\phi_s - \phi_b = (kT/4e)(W/L_D)^2 = \Delta\phi_{sc} \tag{26}$$

where W is the width of the depletion layer, and L_D is the Debye length. The latter was introduced in Chapter 2 in connection with the potential distribution around a spherical colloidal particle in solution:

$$L_D = (\epsilon_0 \epsilon kT/2e^2 \, ^1N)^{1/2} \tag{27}$$

^1N is the concentration of ionized dopants, expressed in number of ions per cubic centimeter. From Equation 26 it is apparent that for a given potential drop within the semiconductor, the width of the space charge layer is proportional to the Debye length, i.e., it decreases with the square root of dopant concentration.

In Figure 11 we presented the position of the band edges of a variety of semiconductors. These results were obtained from measuring the differential capacity of the semiconductor-electrolyte junction. The semiconductor is subjected to reverse bias, and the differential capacity is determined as a function of the applied potential. The space charge capacity of the semiconductor (C_{sc}) is in series with that of the Helmholtz layer (C_H) present at the electrolyte side of the interface. Thus, the total capacity is given by

$$1/C = 1/C_{sc} + 1/C_H \tag{28}$$

Applying a parallel place capacitor model to the space charge region, one can express C_{sc} in terms of the depletion layer width:

$$C_{sc} = \epsilon \cdot \epsilon_0/W \tag{29}$$

Expressing W by Equation 26, one obtains

$$1/C_{sc}^2 = \frac{2}{\epsilon\epsilon_0 e^1N} (\Delta\phi_{sc}) \tag{30}$$

A similar equation was derived by Schottky:

$$1/C_{sc}^2 = \frac{2}{\epsilon\epsilon_0 e^1N} \left\{ |\Delta\phi_{sc}| - \frac{kT}{e^-} \right\} \tag{31}$$

and this is frequently used in the evaluation of the capacity measurements. It is assumed that the condition $C_H \gg C_{sc}$ applies and, therefore, that the measured capacity is that of the space charge layer. Since the capacity of the Helmholtz layer is relatively large ($>10\mu F/cm^2$) and does not vary significantly with the applied potential, this assumption is reasonable. Note, however, that it breaks down for semiconductors with large dielectric constants and high doping levels where C_H and C_{sc} are of comparable magnitude.[56] Also, this condition is not fulfilled for semiconductors having a high density of surface states, as discussed in Chapter 1, Section IV.C.2. A plot of the reciprocal capacity against the applied voltage (Mott-Schottky plot) gives a straight line in the depletion region, and this is extrapolated to $1/C_{sc} = 0$ in order to evaluate the flat band potential (Figure 14). From the slope of the straight line, the concentration of ionized dopant can be derived.

Meanwhile, flat band potentials have been determined for a large number of semicon-

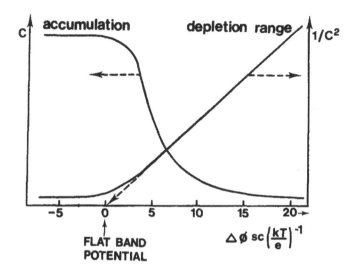

FIGURE 14.   Differential capacity and Mott-Schottky plot for a depletion
layer as a function of the voltage drop in the space charge layer.

ductors under different experimental conditions.[57] Apart from the semiconductor material
these depend on the nature and composition of the electrolyte. Thus, specific adsorption of
ions can shift the flat band position significantly. Of particular importance in this respect
are potential determining ions. Potential determining ions play the same role in semiconductors as do electrons in metal electrodes. Changing the concentration changes the potential
drop across the electrolytic double layer.

Consider the case of an oxide semiconductor particle dispersed in water or an oxide
electrode in contact with an aqueous electrolyte to which no external bias is applied. For
oxides, H^+ and OH^- are potential determining ions. Using the same thermodynamic arguments developed in Chapter 1, the potential difference between the oxide surface and
solution is given by

$$(\phi_s^{ox} - \phi^{sol}) = \frac{(\mu_{H^+}^*)^{ox} - (\mu_{H^+}^*)^{sol}}{F} + \frac{RT}{F} \ln \frac{(a_{H^+})^{sol}}{(a_{H^+})^{ox}} \tag{32}$$

where $(a_{H^+})^{ox}$ and $(a_{H^+})^{sol}$ are the activities of protons on the surface in the plane of adsorption
of the potential determining ions and in solution, respectively. In the case where the proton
activity on the surface does not change with pH, Equation 28 simplifies to

$$(\phi_s^{ox} - \phi^{sol}) = \text{const} - 0.059 \text{ pH} \tag{33}$$

Thus, the potential difference between the semiconductor and solution changes at room
temperature by 59 mV for every pH unit. Criteria for such Nernstian behavior have been
discussed by Hunter.[58] The surface of oxides is composed of amphoteric sites which can
become charged either positively or negatively:

$$M\text{-}O^- + H^+ \leftrightarrows MOH \tag{34}$$

$$M\text{-}OH + H^+ \leftrightarrows M\text{-}OH_2^+ \tag{35}$$

Such a two-site model gives for the change in the surface potential with pH the expression

$$d(\phi_s^{ox} - \phi^{sol})/dpH = \left(-\frac{2.3\,kT}{e}\right) - \left(\frac{kT}{2N_s e^2}\right)\left(\frac{1}{\theta_O}\right)\left(\frac{d\sigma_O}{dpH}\right) \tag{36}$$

where the first term has a value of 59 mV at room temperature and the second gives the deviation from Nernstian behavior. The parameters that influence the magnitude of the deviation are the density of ionizable groups ($N_s[cm^{-2}]$); the change of the surface charge (σ_O) with pH; and θ_O, which depends on the difference in the equilibrium constants for the surface protonation reactions (Equations 34 and 35):

$$\theta_O = (10^{[pK(34) - pK(35)]/2} + 2)^{-1} \tag{37}$$

θ_o reaches a maximum value of 0.5 in the case where pK (3.30) − pK (3.31) has a large negative value. This implies that at the point of zero charge the surface concentration of M-OH_2^+ = M-OH^- groups is very high and equals $N_s/2$. In such a situation Nernstian behavior is expected. For titanium dioxide, $\Theta_O = 8.3 \times 10^{-2}$, and $d\sigma_O/pH = 4 \times 10^{-6}$ C per square centimeter. Equation 36 predicts in this case a relatively small deviation (about 7 mV) from Nernstian behavior. Flat band potential measurements performed with TiO$_2$ electrodes showed that at room temperature the band edges shift 59 mV negative for every increasing pH unit:

$$E_{cb} = +0.1 - 0.059 \times pH \text{ (V, NHE)} \tag{38}$$

A similar result was obtained for colloidal TiO$_2$ particles.[29]

$$E_{cb} = -0.1 - 0.059 \times pH \tag{39}$$

Important for the use of colloidal TiO$_2$ in water cleavage devices is the fact that conduction band position is well matched to the hydrogen evolution potential. A similar result was obtained by Ward et al.[59] with pure anatase powder. Thus, over the whole pH domain, the reaction $H^+ + e^- \rightarrow \frac{1}{2}H_2$ has enough driving force to make high rates of hydrogen evolution feasible in the presence of suitable catalysts.

3. Space Charge Layers in Semiconductor Particles

Using the linearized Poisson-Boltzmann equation, Albery and Bartlett[60] have derived the potential distribution in a spherical semiconductor particle. They obtain for the difference in the potential at the center ($r = 0$) and at a distance r the relation

$$\Delta\phi = \left(\frac{kT}{6e}\right)\left(\frac{r - (r_O - W)}{L_D}\right)^2\left(1 + \frac{2(r_O - W)}{r}\right) \tag{40}$$

A graphic illustration of the potential distribution for an n-type semiconductor particle, which is at equilibrium with a solution containing a redox system for which the Fermi level is E_F, is shown in Figure 15. We discuss here two limiting cases of Equation 40 which are particularly important for light-induced electron transfer in semiconductor dispersions. If the size of the particles is much larger than the depletion layer width ($r_O \gg W$), the condition $r_O \simeq r$ holds within the depletion layer, and Equation 40 simplifies to

$$\Delta\phi = \frac{kT}{2e}\left(\frac{r - (r_O - W)}{L_D}\right)^2 \tag{41}$$

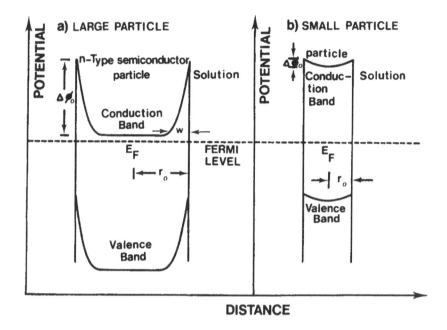

DISTANCE

FIGURE 15. Space charge layer formation in a large and small semiconductor particle in equilibrium with a solution redox system for which the Fermi level is E_F. Of the charge carriers, the
small particle depletes almost completely. Hence, the Fermi potential is located approximately in
the middle of the band gap and band bending is negligibly small.

For $r = r_0$ one obtains

$$\Delta\phi_O = \frac{kT}{2e}\left(\frac{W}{L_D}\right)^2 \tag{42}$$

where $\Delta\phi_O$ corresponds to the total potential drop within the semiconductor particle. Note
that Equation 42 is identical to Equation 26, which was derived for semiconductor electrodes.
Apparently, for large particles, the depletion layer is the same as that for planar electrodes.

For very small (colloidal) semiconductor particles, the condition $W = r_0$ holds, and
Equation 40 reduces to

$$\Delta\phi = \frac{kT}{6e}\left(\frac{r}{L_D}\right)^2 \tag{43}$$

For the total potential drop within the particle one obtains

$$\Delta\phi_O = \frac{kT}{6e}\left(\frac{r_O}{L_D}\right)^2 \tag{44}$$

From this equation and Figure 15 it is apparent that the electrical field in colloidal
semiconductors is usually small and that high dopant levels are required to produce a
significant potential difference between the surface and the center of the particle. For example, in order to obtain a 50-mV potential drop in a colloidal TiO_2 particle with $r_O = 6$
nm, a concentration of 5×10^{19} cm^{-3} of ionized donor impurities is necessary. Undoped
TiO_2 colloids have a much smaller carrier concentration, and the band bending within the
particles is therefore negligibly small. If majority carriers depleted from a colloidal semi-

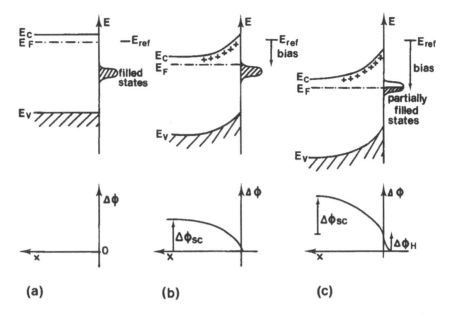

FIGURE 16. Fermi level pinning at an n-type semiconductor with filled electronic surface states. (a) Flat band potential; (b) the electrode is submitted to reverse bias and the Fermi level has attained the energy of the surface states; and (c) ionization of the surface states changes the charge in the Helmholtz layer leading to a downward shift of the band edges at the interface.

conductor in solution and the particles are too small to develop a space charge layer, the electrical potential difference resulting from the transfer of a charge from the semiconductor to the solution must drop in the Helmholtz layer, neglecting diffuse layer contributions. As a consequence, the position of the band edges of the semiconductor particle will shift. The same considerations hold for the transfer of minority carriers formed by photoexcitation of the semiconductor particle. For example if, after photoexcitation of a colloidal n-type particle, holes are transferred rapidly to an acceptor in solution while electrons remain in the particle, a negative shift in the conduction band edge at the surface will take place.

4. Fermi Level Pinning

We have assumed so far that the voltage resulting from the depletion of majority charge carriers from the semiconductor into the electrolyte drops entirely across the space charge region and has no effect on the potential distribution within the Helmholtz layer. This assumption is no longer valid if there is a significant concentration of electronic states at the surface, whose energetic position is within the band gap of the semiconductor.

The latter case is analyzed in Figure 16, where we consider an n-type semiconductor with midband gap surface states centered around the energy E_{ss}. If the Fermi level is above the energy E_{ss}, these states are filled, and the applied voltage will drop across the semiconductor space charge layer. However, upon increasing the anodic bias, the Fermi level will reach the energy of the surface states where it gets pinned until all these states have been discharged. This discharge changes the potential drop across the Helmholtz layer, leading to an anodic shift ($\Delta\phi_H$) of the band edges of the semiconductor:

$$\Delta\phi_H' = e \cdot N_{ss}/C_H \qquad (45)$$

where N_{ss} is the density of surface states. For a Helmholtz capacity of 10 $\mu F/cm^2$, the density of surface states must be around 6×10^{13} cm^{-2} in order to obtain a shift of $+1$ V in the

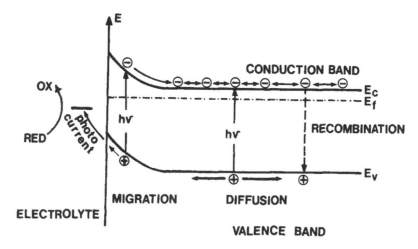

FIGURE 17. Light-induced charge separation assisted by the local electrostatic field present in the depletion layer of a semiconductor electrode in contact with an electrolyte.

band edges. This is a relatively large density of surface states, since it corresponds to approximately 5% of a monolayer of surface atoms.

5. Light-Induced Charge Separation

The depletion layer at the semiconductor-solution interface, sometimes referred to as Schottky barrier, plays an important role in light-induced charge separation. The local electrostatic field present in the space charge layer serves to separate the electron-hole pairs generated by illumination of the semiconductor (Figure 17). For n-type materials, the direction of the field is such that holes migrate to the surface, where they undergo a chemical reaction, while the electrons drift through the bulk to the back contact of the semiconductor and subsequently through the external circuit to the counterelectrode. Charge carriers which are photogenerated in the field-free space of the semiconductor can also contribute to the photocurrent. In solids with low defect concentration, the lifetime of the electron-hole pairs is long enough to allow for some of the minority carriers to diffuse to the depletion layer before they undergo recombination.

Neglecting recombination within the space charge layer, Gärtner[61] obtained for the photocurrent density the expression

$$i_{ph} = eI_0(1 - \exp(-\alpha W)/(1 + \alpha L)) \tag{46}$$

where I_0 is the incident light intensity, α is the reciprocal absorption length, and L is the minority carrier diffusion length. The latter is equal to the square root of the product of the diffusion coefficient and the mean carrier lifetime, $L = (D \times \tau)^{1/2}$. The quantum efficiency obtained from Equation 41 by linearizing the exponential is

$$\eta = (L + W)/(1/\alpha + L) \tag{47}$$

This equation predicts that the quantum yield of a photocurrent generation should increase steeply as soon as the potential of the semiconductor electrode departs from the flat band value and a depletion layer is formed. This is illustrated schematically in Figure 18 for a semiconductor with a band gap of approximately 1.5 V which is n-doped in Figure 18A and p-doped in Figure 18B. The experimental data usually deviate from this behavior in that the rise in the photocurrent is less steep than predicted by the Gärtner model. In addition,

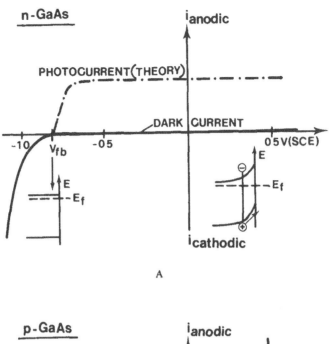

A

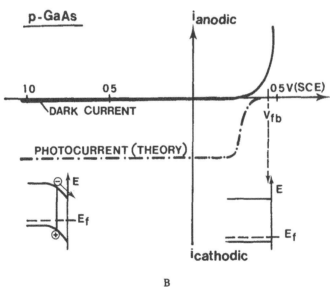

B

FIGURE 18.   Photocurrent voltage curves predicted from the Gärtner model for a semiconductor, i.e., GaAS, with a band gap of approximately 1.5 eV. (A) n-Type and (B) p-type. The saturation current corresponds to a quantum yield, corrected for reflection losses, of close to 100%.

the photocurrent onset is frequently shifted from V_{fb} to a potential where the band bending is already several hundred millivolts. This behavior can be accounted for by a recombination of charge carriers (Schockley-Read mechanism) which takes place mainly at the surface. Also, the Gärtner model assumes that the reaction of the minority carrier with solutes is so fast that it has no effect on the photocurrent. However, there is evidence that this interfacial electron or hole transfer can become an important factor, controlling the efficiency of photocurrent generation under near flat band conditions.

In the case of colloidal semiconductors, the band bending is small, and charge separation occurs via diffusion. The absorption of light leads to the generation of electron-hole pairs in the particle which are oriented in a spatially random fashion along the optical path. These charge carriers subsequently recombine or diffuse to the surface where they undergo chemical reactions with suitable solutes or catalysts deposited onto the surface of the particles. Applying the random walk model developed in Chapter 2 for the diffusion of guest molecules in micelles in order to describe the motion of the charge carriers, one obtains for the average transit time from the interior of the particle to the surface the expression

$$\tau_d = r_0^2/\pi^2 D \tag{48}$$

For colloidal semiconductors τ_d is at most a few picoseconds. Thus, for colloidal TiO_2 ($D_{e-} = 2 \times 10^{-2}$ cm^2/sec) with a radius of 6 nm the average transit time of the electron is 3 psec.

In the presence of a depletion layer, the transit time of the minority carriers is further reduced, since in this case the Debye length must be used in Equation 48 instead of the particle radius.[60] However, as pointed out by Curran and Lamouche,[62] the potential difference between the particle surface and interior must be at least 50 mV in order for migration to dominate over diffusion.

It should be noted that the random walk model breaks down for particles exhibiting quantum size effects. Here, the wave function of the charge carrier spreads over the whole semiconductor cluster, and it does not have to undergo diffusional displacement to accomplish reactions with species present at the surface.

Since in colloidal semiconductors the diffusion of charge carriers from the interior to the particle surface can occur more rapidly than the recombination, it is feasible to obtain quantum yields for photoredox processes approaching unity. Whether such high efficiencies can really be achieved depends on the rapid removal of at least one type of charge carrier, i.e., either electrons or holes, upon the arrival at the interface. This underlines the important role played by the interfacial charge transfer kinetics, which we shall treat in further detail below.

6. Trapped vs. Free Charge Carriers in Small Semiconductor Particles

A final point to be discussed in this section concerns the distribution of charge carriers between the conduction or valence band and trapping states. Heterogeneous photochemical charge transfer reactions are often performed with polycrystalline semiconductors containing trapping sites in the bulk and at the surface. The question which will be addressed now concerns the minimum trap depth required for effective direct trapping of a mobile charge carrier. There is some confusion with respect to this question in the literature. The depth of a trap, for example, for a conduction band electron is usually expressed in terms of the difference in internal energy of the electron at the bottom of the conduction band and the trapping level. Defined in this way the trap depth is not necessarily a good indication of whether electron trapping will be effective or not, since the entropy contribution is neglected. As the number of translational energy states available to the free electron in the conduction band is much higher than the number of traps, the electron trapping entails a significant decrease in entropy. This must be taken into account when considering trapping probabilities.

Consider a small semiconductor particle of volume V containing n_t trapping sites all having the same depth, i.e., the same internal energy $-E_t$. (The internal energy of the bottom of the conduction band is used as a reference and is set equal to zero: $E_{cb} = 0$). If electronic equilibrium is established, the chemical potentials of trapped and free electrons are equal and correspond to the Fermi level of the semiconductor particle:

$$\mu_t(e^-) = \mu(e^-) = E_f \tag{49}$$

We use this equilibrium condition to derive the distribution of electrons between the trapping sites and the conduction band. For trapped electrons,

$$\mu_t = \left(\frac{\partial F}{\partial N_t} \right)_{V,T} \tag{50}$$

where N_t is the number of trapped electrons, and F_t is the Helmholtz free energy defined as

$$F_t = U_t - TS_t \tag{51}$$

The internal energy of the ensemble of trapped electrons is given by

$$U_t = -N_t E_t \tag{52}$$

If there are n_t traps in the semiconductor particles, all having the same energy, only the configurational part of the entropy needs to be considered, which is given by

$$S_t = k \ln \left(2^{N_t} \frac{n_t!}{(n_t - N_t)! N_t!} \right) \tag{53}$$

where the factor 2^{N_t} arises from the two possible spin orientations of the trapped electron. Using Stirling's approximation, the entropy becomes

$$S_t = k[N_t \ln(2/N_t) + n_t \ln n_t - (n_t - N_t) \ln(n_t - N_t)] \tag{54}$$

Inserting Equations 54 and 52 into Equation 51 and differentiating gives

$$\mu_t = -E_t + kT \ln[N_t/2(n_t - N_t)] \tag{55}$$

The chemical potential of free conduction band electrons is given by Equation 23:

$$\mu_f = E_f = kT \ln(N_{cb}/n_{cb}) \tag{56}$$

Inserting these two expressions in the equilibrium condition (Equation 49), one obtains for the ratio of free and trapped electrons

$$N_{cb}/N_t = \frac{n_{cb}}{2(n_t - N_t)} \exp(-E_t/kT) \tag{57}$$

The predictions of Equation 58 are illustrated by considering the trap depth necessary for trapping 50% of the conduction band electrons. In this case $N_{cb}/N_t = 1$, and Equation 58 reduces to

$$E_t/kT = \ln[n_{cb}/2(n_t - N_t)] \tag{58}$$

As an example, we consider colloidal TiO_2 particles of a 6-nm radius for which the number of accessible conduction band states is $n_{cb} = 3600$. If the number of available traps $(n_t - N_t)$ is 100, the value of E_t calculated from Equation 59 is 0.072 V, which corresponds to about 3 times the mean thermal energy at 298°K. More detailed calculations have been published by Mitchell[63] for silver halide microcrystals.

III. EXPERIMENTAL METHODS

We distinguish stationary from pulsed methods. For semiconductor electrodes, the stationary method consists of the measurement of the photocurrent as a function of applied bias. It is difficult to derive from such measurements the rate constant for the interfacial charge transfer step. It was shown in the previous section that the photocurrent is influenced by a charge carrier recombination in the bulk and at the surface. Moreover, the heterogeneous charge transfer is often preceded by the trapping of the minority carriers in surface states. Thus, the overall photocurrent contains contributions from several processes which are difficult to discern. Although the individual rate constants for the interfacial charge transfer cannot be derived from such measurements, relative rates for minority carrier reactions with different scavengers present in the electrolyte can readily be determined this way.[64] In addition, the efficiency of various redox species in suppressing the photocorrosion of semiconductor electrodes can be determined from such photocurrent measurements.

Stationary methods have also been applied to suspensions of semiconductor particles[65,66] and colloids.[67] These experiments employ a conventional electrochemical cell, the semiconductor particles being dispersed in the compartment of the working electrode. Excitation by band gap irradiation produces electron-hole pairs in the particles. The electrons or holes are subsequently transferred to a suitable relay molecule dissolved in the electrolyte. This in turn transports the charge to the collector electrode, producing a current which is measured as a function of the applied potential. These investigations yield valuable information on the position of the Fermi level within the semiconductor particle and on the kinetic features of heterogeneous electron-transfer reactions involving colloidal semiconductors on electrodes.

With colloidal semiconductors, the diffusion of the particles is rapid enough to allow for the experiments to be performed in the absence of relay compounds. In this case, the exchange of mobile charge carriers between the semiconductor colloid and the collector electrode can be directly measured, yielding information on the rate of the interfacial redox reaction as well as the energetic position of the electronic energy bands within the particle.[67]

Perone et al.[68] have developed pulsed laser techniques for the study of light-induced processes on semiconductor electrodes. Transient coulostatic photopotentials could be studied with about a 10-nsec time resolution. It was suggested that the photopotential decay arises from space charge relaxation. A similar technique was developed by Herzion et al.[69] and Gottesfeld and Feldberg.[70] Following their work, several authors investigated pulsed laser-induced photopotentials or photocurrents on different semiconductors such as TiO_2,[71] WSe_2,[72] Fe_2O_3,[73] PtS_2,[74] and CdSe.[75] Laser-induced transients at the rhodamine B-SnO_2 interface resulting from spectral sensitization were also studied.[76] More recently, Bitterling and Willig[77] have succeeded in obtaining a very high time revolution by using picosecond laser pulse excitation to study recombination and interfacial charge transfer with semiconductor electrodes.

A direct access to the analysis of heterogeneous charge transfer reaction is provided by laser excitation of solutions of semiconductor particles, in particular colloids. The latter are very attractive for time-resolved studies, since they yield transparent solutions. This allows for the ready analysis of the extremely fast interfacial electron-transfer events occurring in these systems. Examples will be presented in the following sections which illustrate that the primary events of light-induced charge separation — carrier recombination and trapping as well as the heterogeneous redox events at the semiconductor surface leading to chemical change — are frequently within the picosecond time domain. Only by exploiting the transparent nature of semiconductor sols can such processes be directly monitored.

In Table 3 we summarize the state of the art concerning the time-resolved experimental methods which are currently applied in the analysis of heterogeneous electron-transfer events in semiconductor dispersions. Picosecond laser excitation coupled with kinetic spectroscopy, e.g., UV, visible, or Raman, gives the highest resolution, which is in the 10^{-12} sec time

Table 3
EXPERIMENTAL METHODS: PULSED

Band Gap Excitation with a Short Laser Pulse: Detection of Charge Carriers or
Intermediate Products of Reactions

Particle type	Detection method	Time resolution (sec)
Colloids only	UV visible	
	Absorption	10^{-12}
	Raman	10^{-12}
	Conductivity	10^{-9}
All	Microwave	
	UV visible	10^{-9}
	Reflection	10^{-9}
	Luminescence	10^{-12}

domain. Apart from colloids, the analysis of fast electron-transfer reactions on suspensions of larger particles or on dry powders has made good progress. Time-resolved reflection and microwave loss spectroscopy have so far given the most promising results.[78-81]

IV. DYNAMICS OF CHARGE CARRIER TRAPPING AND RECOMBINATION IN SMALL SEMICONDUCTOR PARTICLES

We shall illustrate the principle of these investigations by reporting studies with colloidal TiO_2 (anatase) particles having a diameter of 120 Å. Irradiation of such colloidal solutions in the presence of a hole scavenger such as polyvinyl alcohol or formate ions results in the accumulation of electrons in the particles. As a result, the solution assumes a beautiful blue color under illumination. It was found that up to 300 electrons can be stored in 1 TiO_2 particle. (A TiO_2 particle of 120-Å size has about 3600 conduction band states. Therefore, at most 10% of the available states are occupied by electrons.) The absorption spectrum of these stored electrons is shown in Figure 19.

The possibility of accumulating charges in a reaction space of minute dimensions is a very important feature of colloidal semiconductor particles. It should be recalled that light-energy conversion processes such as the cleavage of water in hydrogen and oxygen, or the reduction of carbon dioxide involve without exception multielectron-transfer processes. Hence, there is a need for structural organization that provides the possibility of charge storage. In natural photosynthesis, this function is assumed by the plastoquinone assembly in the thylakoid membrane. The colloidal TiO_2 particles investigated here provide an inorganic equivalent for such electron reservoirs.

One might expect that the storage of 300 electrons in a 50-Å-sized particle should produce a large negative shift in potential with respect to the surrounding aqueous solution. However, the TiO_2-water interface is distinguished by a large double layer capacity, which is 85 $\mu F/cm^2$ at the point of zero charge and increases to 200 to 300 $\mu F/cm^2$ in alkaline solution.[82] The negative shift in the potential of the particle during accumulation of 300 electrons is therefore only 25 to 90 mV. This explains why such highly charged particles can be preserved in an anaerobic aqueous environment for at least several weeks without giving rise to significant hydrogen generation. The situation appears to be different with colloidal CdS particles, where laser excitation has been reported to lead to huge negative shifts in the particle potential, resulting in ejection of negative charge carriers to produce hydrated electrons.[83]

The electron spectrum was found to be sensitive to the pH of the solution. Under alkaline conditions the electron absorption is very broad and has a maximum around 800 nm.

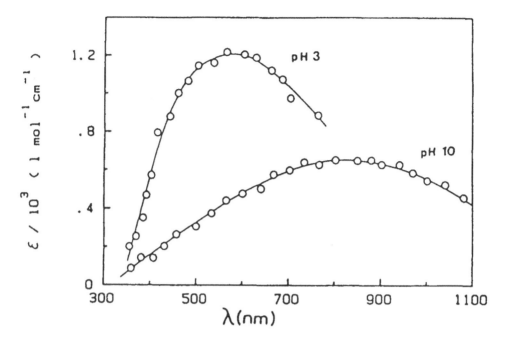

FIGURE 19. Absorption spectrum of conduction band electrons in colloidal TiO_2 particles at pH 3 and pH 10.

Lowering the pH to 3 produced a pronounced blue shift in the spectrum, which under these conditions shows a peak at 620 nm. The sensitivity of the electron absorption to the solution pH would indicate that they are located in the surface region of the particles. This has been confirmed by recent ESR experiments which show that under acidic conditions the electrons are trapped at the TiO_2 surface in the form of Ti^{3+} ions.[84] Using redox titration, we have recently been able to determine the extinction coefficient of the trapped electrons.[85] For the colloidal solutions of pH 3 the extinction coefficient at 600 nm is 1200 mol^{-1} cm^{-1}.

Taking advantage of the characteristic optical absorption of trapped electrons in the colloidal TiO_2 particles, we have recorded the recombination with free and trapped holes in the picosecond-to-microsecond domain.[86] Figure 20 shows the temporal evolution of the transient spectrum after excitation of TiO_2 with a frequency-tripled (353-nm) Nd laser pulse of about 40-psec duration.

In Figure 20 the spectrum of the trapped electron develops within the leading edge of the laser pulse, indicating that the trapping time of the electron is less than 40 psec. This is in accordance with the predictions of Equation 48. The electron absorption decay is due to recombination with valence band holes:

$$TiO_2(e_{tr}^- + h^+) \xrightarrow{k_r} TiO_2 \qquad (59)$$

We have conceived a stochastic model for analyzing the kinetics of this reaction. Since the recombination takes place between a small number of charge carriers restricted to the minute reaction space of a 120-Å-sized colloidal TiO_2 particle, it cannot be treated by conventional homogeneous solution kinetics. The time differential of the probability that a particle contains x electron-hole pairs at time t is given by

$$dP_x(t)/dt = k(x + 1)^2 P_{x+1}(t) - kx^2 P_x(t) \qquad (60)$$

where $x = 0, 1, 2....$ This expression is similar to Equation 16 in Chapter 2 describing intramicellar triplet annihilation.

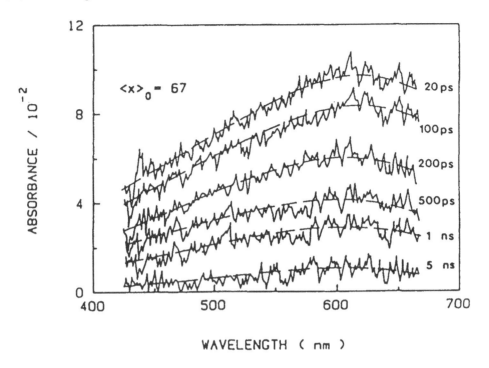

FIGURE 20.   Transient spectrum observed at various time intervals after picosecond excitation of colloidal TiO$_2$. Conditions: (TiO$_2$) = 17 g/ℓ, pH, 2.7; Ar-saturated solution; optical path length, 0.2 cm. Average number of electron-hole pairs present initially in one TiO$_2$ particle is 67.

This system of differential equations is to be solved subject to the condition that the initial distribution of electron-hole pairs over the particles follows Poisson statistics. The average number of pairs present at time t, $\langle x \rangle$ (t), can be calculated by means of the generating function technique,[86] yielding

$$\langle x \rangle(t) = \sum_{n=1}^{\infty} c_n \exp(-n^2 kt) \tag{61}$$

where

$$c_n = 2 \exp(-\langle x \rangle_o)(-1)^n n \sum_{i=n}^{\infty} \frac{\langle x \rangle_o^i}{(n + i)!} \sum_{j=1}^{n} (n - i - j) \tag{62}$$

The parameter $\langle x \rangle_o$ is the average number of pairs present at t = 0.

Two limiting cases of Equation 61 are particularly relevant. When $\langle x \rangle_o$ is very small, Equation 61 becomes a simple exponential, and the electron-hole recombination follows a first-order rate law. Conversely, at a high average initial occupancy of the semiconductor particles by electron-hole pairs, i.e., $\langle x \rangle_o > 30$, Equation 61 approximates to a second-order rate equation:

$$\langle x \rangle(t) = \frac{\langle x \rangle_o}{1 + \langle x \rangle_o kt} \tag{63}$$

In Figure 20 the initial concentration of electron-hole pairs was sufficiently high to allow the evaluation of the recombination process by the second-order rate equation (Equation

63). This analysis gives for the recombination rate coefficient the value 3.2×10^{-11} cm^3 sec^{-1}, corresponding to a lifetime of 30 ns for an electron-hole pair in a colloidal TiO$_2$ particle with a size of 120 Å. From this value one derives a diffusion length of 2.2×10^{-5} cm for the electron.

Experiments carried out at low laser fluence showed that hole trapping at the surface can compete with the recombination. In aqueous dispersions of oxide semiconductors such as TiO$_2$, hydroxide ions are present at the surface which can trap positive holes. Note that there are two types of OH$^-$ groups: basic and acidic.[87] The former are attached to one Ti^{4+} ion, while the latter bridge two Ti^{4+} sites. Since the electron density is higher on the basic OH$^-$, it is expected that these act as the hole traps:

$$h^+ + \underset{|\ |}{\overset{OH}{\underset{|}{\ }}}\ Ti\ \underset{|\ |}{\overset{O}{\ }}\ Ti\ \overset{OH}{\underset{|}{\ }} \rightarrow \overset{\dot{O}}{\underset{|}{\ }}\ Ti\ \overset{O}{\ }\ Ti\ \overset{OH}{\underset{|}{\ }} + H^+ \tag{64}$$

The laser photolysis experiments with colloidal TiO$_2$ yield a rate constant of 4×10^5 sec^{-1} for this reaction. The standard free energy change associated with Reaction 64 has recently been derived from luminescence experiments as about -1.4 eV.[88] Thus, basic hydroxide groups at the surface of TiO$_2$ can be considered as deep traps for valence band holes. Once the positive hole is trapped at such an OH$^-$ site, we can consider it as a surface bound hydroxyl radical ($\cdot$O$^-$). This species has a surprisingly low cross section for recombination with conduction band electrons ($\sigma \sim 10^{-16}$ cm^2).[89] As a consequence, a small fraction of $\cdot$O$^-$ can escape from recombination with e_{cb}^- and undergo dimerization at the particle surface to form stable titanium peroxide.[90] This explains why irradiation of TiO$_2$ suspensions produces excess electrons in the particles, even in the absence of external electron donors, leading to a downward shift in the isoelectric point.[65] The recombination rate of electron-hole pairs has recently been determined for semiconductor colloids other than TiO$_2$. For example, Nosaka and Fox[91] have derived a value of $(9 \pm 4) \times 10^{-11}$ cm^3 sec^{-1} for the recombination coefficient for CdS particles.

A simple and effective way of influencing the charge carrier recombination dynamics is by introducing suitable dopants in the semiconductor. For example, substitutional doping of TiO$_2$ with Fe^{III} leads to a drastic retardation of the electron-hole recombination rate. This is the origin of the well-known photochromism exhibited by Fe^{III} containing TiO$_2$ particles.[92,93] Under illumination in an inert gas atmosphere, a black-green color is produced which fades very slowly after the light has been turned off. Recent combined EPR and laser photolysis studies with Fe^{III}-doped TiO$_2$ colloids in aqueous solution have helped to elucidate the mechanism of this photochromic effect.[94] Incorporation of Fe^{III} ions in the colloidal particles leads to the formation of substitutional Fe^{III} on Ti^{IV} lattice sites and to charge-compensating oxygen vacancies:

$$Fe_2O_3 \rightarrow 2|Fe^{III}|'_{Ti} + {}^3/_2\,O_O + V_O^{\cdot\cdot} \tag{65}$$

The former act as traps for valence band holes:

$$|Fe^{III}|'_{Ti} + h^+ \rightarrow |Fe^{IV}|_{Ti} \qquad \Delta H^\circ = -0.12 \text{ eV} \tag{66}$$

while the latter scavenge conduction band electrons:

$$V_O^{\cdot\cdot} + e_{cb}^- \rightarrow V_O^{\cdot} \qquad \Delta H^\circ = -0.08 \text{ eV} \tag{67}$$

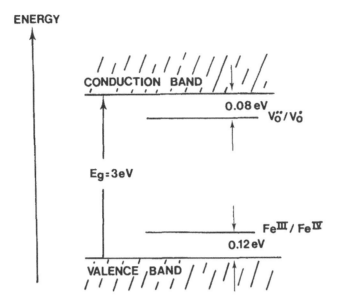

FIGURE 21.   Energetic positions of the Fe^{III}-Fe^{IV} and $V_0^{\cdot\cdot}$-$V_0^{\cdot}$ μ levels with respect to the band edges of TiO_2.

Figure 21 illustrates the energetic position of the two trapping levels with respect to the band edges of TiO_2. These traps are sufficiently deep to insure immobilization of the electrons and holes on the trapping sites at room temperature. Therefore, the recombination involves tunneling of the charge carriers through the solid. We have already discussed the dynamics of such processes in the context of radiative and radiationless recombination of trapped electron-hole pairs in Section II.B.4. The kinetic description is given by Equation 20. Recall that the rate constant for recombination decreases exponentially with the distance R separating the trapped electron from the trapped hole (Figure 10). As a consequence, charge carriers that are relatively far apart (R > 30 Å) recombine slowly, the survival time being in the domain of several hours.

V. DYNAMICS OF INTERFACIAL CHARGE TRANSFER PROCESSES

Having dealt in detail with the formation, motion, and storage of electrons and holes within ultrafine semiconductor particles, we address now the question of interfacial charge transfer. This is a crucial step in the overall process. We need to remove either the electrons or the holes or both quickly from the particles in order to avoid charge recombination and to achieve good efficiencies. A strategy frequently employed is to deposit a catalyst at the particle surface which can act as an electron or hole trap (Figure 22). Alternatively, the electrons or holes might react with an appropriate acceptor molecule in solution. In the latter case, the overall reaction is composed of two steps:

1.   Encounter complex formation of the electron (or hole) acceptor with the semiconductor particle. The rate of this process is diffusion limited and hence determined by the viscosity of the medium and the radius of the reactants.
2.   Interfacial electron transfer. This electrochemical step involves a Faradayic current across the semiconductor-solution interface and is characterized by the rate parameter k_{et} (units, centimeters per second).

(1) EXCITATION & CHARGE CARRIER
    DIFFUSION TO THE INTERFACE

(2) TRAPPING BY SURFACE ADSORBED
    CATALYST OR ACCEPTOR

(3) ENCOUNTER WITH ACCEPTOR
    VIA DIFFUSION

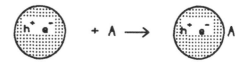

(4) INTERFACIAL CHARGE TRANSFER

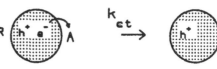

FIGURE 22. Elementary steps in the heterogeneous electron transfer induced by light from a semiconductor particle to an acceptor or a catalyst.

For the kinetic treatment of this reaction sequence, we consider the radial diffusion field around one semiconductor particle. The flux of the electron donor (or acceptor) is

$$J = D \frac{dc}{dr} \qquad (68)$$

and the temporal change of concentration within the diffusion field is

$$\frac{dc}{dt} = D\left(\frac{d^2c}{dr^2} - \frac{2}{r}\frac{dc}{dr}\right) \qquad (69)$$

where r is the distance from the center of the particle. Assuming the diffusion field to be stationary, $dc/dt = 0$, one obtains

$$\frac{d^2c}{dr^2} - \frac{2}{r}\frac{dc}{dr} = 0 \qquad (70)$$

which when integrated within the limits $r = R$ and $r = \infty$ gives

$$\left(\frac{dc}{dr}\right)_{r=R} = \frac{c_b - c_R}{R} \qquad (71)$$

where c_b is the bulk concentration of the electron donor (or acceptor). At the surface, the flux under stationary conditions corresponds to the rate of the heterogeneous electron-transfer reaction:

$$D\left(\frac{dc}{dr}\right)_{r=R} = k_{ct} \cdot c_R \qquad (72)$$

From Equations 71 and 72 one derives for the concentration of the electroactive species at the particle surface

$$c_R = c_b \Big/ \left(1 + \frac{R}{D} k_{ct}\right) \tag{73}$$

and for the flux

$$J = c_b/(1/k_{ct} + R/D) \tag{74}$$

The rate constant for the bimolecular reaction is obtained by multiplying the flux by $4\pi R^2$ and dividing by c_b:

$$\frac{1}{k_{obs}} = \frac{1}{4\pi R^2} \left(\frac{1}{k_{ct}} + \frac{R}{D}\right) \tag{75}$$

where R is the sum of the radii of the semiconductor particle and electron (or hole) acceptor, and D is the sum of the respective diffusion coefficient. A similar equation was derived by Albery and Bartlett[95] to describe the reaction of colloidal metal particles with electron donors or acceptors in solution.

Equation 75 establishes an important link between k_{obs}, the familiar rate constant for a second-order reaction, and the electrochemical rate parameter k_{ct}. The structure of this equation suggests two limiting cases:

1. Heterogeneous charge transfer is rate determining and much slower than diffusion ($k_{ct} \ll D/R$). In this case Equation 75 reduces to

$$k_{obs} = 4\pi R^2 k_{ct} \tag{76}$$

2. Heterogeneous charge transfer is faster than diffusion, which controls the overall reaction rate ($k_{ct} \ll D/R$). In this case we obtain from Equation 75 the well-known Smoluchowski expression:

$$k_{obs} = 4\pi DR \tag{77}$$

Note, finally, that the diffusion effects can be eliminated by adsorption or chemical fixation of the acceptor at the semiconductor particle surface. Here, the simple relation

$$k_{ct} = \delta/\tau_{ct} \tag{78}$$

holds, where τ_{ct} is the average time required for the charge carrier to tunnel across the interface, and δ is the reaction layer thickness.

In the following, we illustrate how heterogeneous rate constants for electron transfer from colloidal semiconductors to acceptors in solution can be determined experimentally. An instructive example is the reduction of cobalticinium dicarboxylate $[Co(CpCOO^-)_2]^-$:[85]

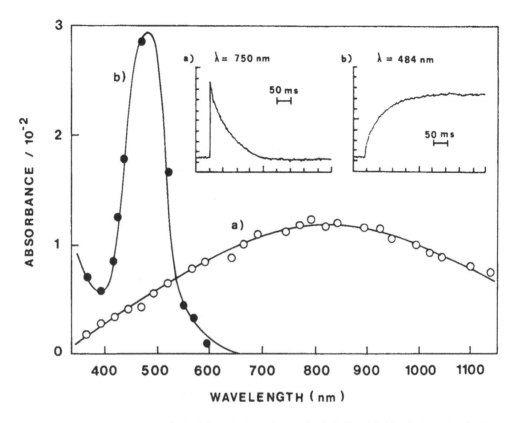

FIGURE 23. Transient spectra obtained from the laser photolysis of alkaline (pH 10) solutions of colloidal TiO$_2$ (0.5 g/ℓ) in the presence of 5 × 10^{-4} Co (CpCOO) $_2^-$ (concentration of PVA is 0.5 g/ℓ). (○, spectrum about 10 μsec after the laser pulse; ●, spectrum 400 msec after the laser pulse.) Insert shows temporal evolution of the absorbance at (a) 750 and (b) 484 nm.

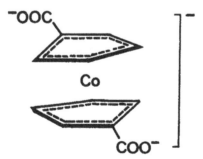

[Co(CpCOO$^-$)$_2$]$^-$ has shown great promise as a relay compound[96] for hydrogen generation in illuminated chloroplast suspensions[97] or in regenerative photoelectrochemical cells based on p-InP electrodes.[98] Advantages of this mediator are high chemical stability and relatively weak visible light absorption in both oxidized and reduced form. In neutral or alkaline water, reversible one-electron reduction occurs[99] at −0.63 V (NHE), rendering the reduction of water to H$_2$ or of CO$_2$ to formiate thermodynamically feasible.

Figure 23 shows absorption spectra obtained from the laser photolysis of alkaline (pH 10) solutions of colloidal TiO$_2$ (0.5 g/ℓ) in the presence of 5 × 10^{-4} M cobalticinium dicarboxylate. Inserted are two oscillograms illustrating the temporal behavior of the optical density at 750 and 484 nm. Immediately after the laser flash the transient spectrum has a maximum at 780 nm and is identical with that shown in Figure 19 for localized electrons in TiO_2 particles. The decay of e_{cb}^- (TiO_2) is exponential and matches the growth of the 484-nm absorbance where reduced cobalticinium dicarboxylate has an absorption maximum. This shows that the process observed in Figure 23 follows the stoichiometry

$$e_{cb}^-(TiO_2) + [Co(CpCOO^-)_2]^- \rightarrow [Co(CpCOO^-)_2]^{2-}$$

$$\lambda_{max}, 780 \text{ nm} \qquad\qquad\qquad \lambda_{max}, 484 \text{ nm} \qquad\qquad (79)$$

As expected, the rate law was found to be pseudo-first order with respect to cobalticinium dicarboxylate concentration, and a second-order rate constant of 4×10^4 mol^{-1} sec^{-1} was evaluated from the kinetic analysis. This is far below the limit imposed by diffusion, which for TiO_2 particles with a 50-Å radius is calculated as 5×10^{10} mol^{-1} sec^{-1}. Therefore, the reaction is controlled by the heterogeneous electron transfer at the particle surface, and we can apply Equation 77 to calculate k_{ct}. The value derived is $k_{ct} = 2 \times 10^{-5}$ cm sec^{-1}, indicating a relatively slow rate for the interfacial redox reaction.

This example illustrates how the characteristic absorption of charge carriers in semiconductor particles can be used to monitor directly interfacial electron-transfer reactions. This opens up the way to determining heterogeneous rate constants for fast charge transfer from the semiconductor to any reactants or catalysts present in solution or deposited on the semiconductor interface. Thus, laser excitation of colloidal semiconductors combined with fast kinetic spectroscopy offers a very useful complement to the application of the conventional electrochemical techniques, which achieve only a relatively low time resolution for probing electron-transfer events at the semiconductor-solution interface.

It was pointed out in Section II.C.1 that the band position of oxide semiconductors in aqueous solution is pH dependent. Increasing the pH by one unit shifts the potential of the valence and conduction band by about 60 mV in the negative direction. One can take advantage of this effect in order to vary the driving force for heterogeneous charge transfer from the semiconductor to a given redox species in solution. Changing the pH changes the overvoltage of the reaction and hence the electron-transfer rate. This effect was studied particularly thoroughly for colloidal TiO_2 and viologen-type acceptors[29,100-104] of the general structure

where

		E° (NHE)
$R_1 = R_2 = CH_3$	Methylviologen (MV^{2+})	-0.44 V
$R_1 = R_2 = (CH_2)_3\text{-}SO_3^-$	Propylviologen sulfonate (PVS)	-0.38 V
$R_1 = CH_3, R_2 = (CH_2)_{13}\text{-}CH_3$	$C_{14}MV^{2+}$	-0.44 V

and the dimeric viologen:[105]

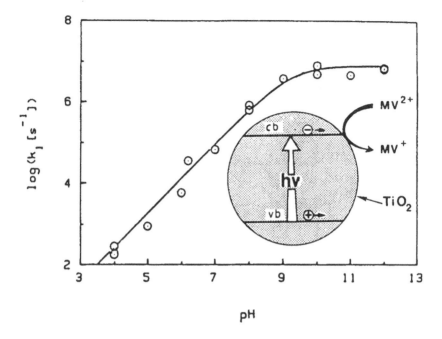

FIGURE 24. Kinetics of reduction of methyl viologen (MV^{2+}) by conduction band electrons of colloidal TiO_2. The observed rate constant for MV^+ formation, derived from 347.1-nm laser photolysis experiments, is plotted as a function of pH. The solid line represents a computer fit with Equation 82 using $\alpha = 0.84$ and $k_{ct}^{\circ} = 10^{-2}$ cm/sec.

Figure 24 shows data obtained from the laser photolysis of colloidal TiO_2 in the presence of MV^{2+} as conduction band electron acceptor. The logarithm of the second-order rate constant for electron transfer is plotted as a function of the solution pH. The k_{obs} values were determined by monitoring the growth of the 602-nm absorption of the viologen cation radicals after excitation of the TiO_2 colloid. A striking 10^5-fold increase in the rate constant is obtained when the solution pH is changed from 4 to 10, the slope of the straight line being 0.74. The log k_{obs} vs. pH plot is linear in this domain. However, the curve bends sharply at higher alkalinity, attaining a limit of 5×10^{10} mol^{-1} sec^{-1}.

For the analysis of these results, we define the standard driving force or standard overvoltage (see Chapter 1) for the interfacial electron transfer as

$$\eta^{\circ} = E_{cb} - E^{\circ}(A/A^-) \tag{80}$$

Furthermore, we use a linear free energy relation, the Tafel equation, in order to express the rate constant for electron transfer as a function of the overvoltage:

$$k_{ct} = k_{ct}^{\circ} \exp(-\alpha\eta^{\circ}F/RT) \tag{81}$$

where k_{ct}° is the specific rate at zero driving force under standard conditions. From Equations 39, 75, 80, and 81, one obtains

$$\frac{1}{k_{obs}} = \frac{1}{4\pi R^2} \left(\frac{1}{k_{ct}^{\circ} 10^{(\alpha[0.11 + 0.059\,pH + E^{\circ}(A/A^-)]F/RT)}} + \frac{r}{D} \right) \tag{82}$$

The solid line in Figure 24 is a computer plot of this equation using the experimentally determined parameters $r = 60$ Å and $D = 10^{-5}$ cm^2/sec and the standard redox potential

Table 4

**HETEROGENEOUS RATE CONSTANTS
AND TRANSFER COEFFICIENTS FOR
THE REDUCTION OF ACCEPTORS BY
CONDUCTION BAND ELECTRONS OF
COLLOIDAL TiO₂ DERIVED BY LASER
PHOTOLYSIS**

Acceptor	k_{ct}° (cm/sec)	α	Ref.
MV^{2+}	10^{-2a}	0.85	105
MV^{2+}	5×10^{-3b}	0.5	100
$C_{14}MV^{2+}$	10^{-3a}	0.78	105
$[Co(C_6H_4COO^-)_2]^-$	2.2×10^{-4a}	0.5	85
O_2	10^{-7b}	0.57	106
Methyl orange		0.62^b	106
$Rh(bipy)_3^{3+}$	0.4^a	0.64	105

[a] TiO_2 colloid prepared via hydrolysis of $TiCl_4$.
[b] TiO_2 colloid prepared via hydrolysis of titanium tetra-isopropoxide.

-0.44 V (NHE) for the MV^{2+}-MV^+ redox couple. The agreement with the results is excellent, supporting the validity of the kinetic model applied. From this curve α is evaluated as 0.85, and $k_{ct}^{\circ} = 0.01$ cm/sec. TiO_2 colloids prepared via the hydrolysis of titanium tetraisopropoxide give a α value of 0.5.[100]

Meanwhile, heterogeneous rate constants and transfer coefficients have been determined for a number of electron acceptors, and some representative examples are given in Table 4. As is apparent from these results, k_{ct} varies over almost 7 orders of magnitude, the highest value being obtained for the reduction of the rhodium complex $Rh(bipy)_3^{3+}$ and the lowest one for that of O_2. Oxygen reduction is kinetically slow on most electrode materials, and hence it is not surprising to find a relatively small value for the electron-transfer rate.

For further interpretation of the results presented in Table 4, we express the reciprocal average time for interfacial electron transfer by the Marcus relation for nonadiabatic electron transfer introduced in Chapter 1, Section III:

$$\frac{1}{\tau_{ct}} = \nu_O \exp[-\beta(d - d_O)] \exp\left(-\frac{(\Delta G^* + \lambda)^2}{4\lambda kT}\right) \qquad (83)$$

The parameter ν_O is the frequency factor reflecting the rate constant of electron transfer for optimal exothermicity when the species are in contact ($d = d_O$). Assuming that at room temperature, ν_O has a value of about $10^{13.3}$ sec^{-1} [107] and that interfacial charge transfer occurs over a distance of $d - d_O = 5$ Å, one obtains from Equation 83 $k_1 = 5 \times 10^{10}$ sec^{-1} when $\beta = 1.2$ Å^{-1} and the driving force ($-\Delta G^* = \lambda$) is optimal. If $\Delta G^* = 0$ and the reorganization energy $\lambda = 0.5$ eV, k_1 equals 3.4×10^8 sec^{-1}. Since the reaction layer has a thickness of about $1/\beta \approx 0.8$ Å, the heterogeneous rate constant at zero driving force $k_{ct}^{\circ} \approx 2.7$ cm/sec is obtained from Equation 78.

The k_{ct}° values reported in Table 4 are smaller than this limit, indicating $\lambda > 0.5$ eV and/or $d > 5$ Å for the acceptors investigated. It should be noted that very high electron-transfer rates ($k_1 > 10^8$ sec^{-1}) were obtained in cases where the acceptor adheres strongly to the surface of the semiconductor. Examples are the one-electron reduction of DV^{4+} adsorbed onto colloidal TiO_2, which is completed within 50 psec after laser excitation of the particles, and the reduction of MV^{2+} by conduction band electrons of Cds.[17,29,34,35,109,110]

The α values reported in Table 4 range from 0.5 to 0.85. Caution should be applied in the interpretation of these values. The kinetic treatment of interfacial electron-transfer reactions presented so far neglects coulombic effects in the diffusional approach of semiconductor particle and acceptor. Therefore, Equation 82 is only valid when electrostatic work terms are negligible, i.e., the acceptor and/or particle is uncharged, or the ionic strength is high. In general, Equation 82 must be corrected[104] to allow for the variation in electrostatic attraction as the pH changes. For example, in the case of the reduction of MV^{2+} by conduction band electrons of colloidal TiO_2, Brown and Darwent[104] found empirically:

$$\log k_{ct} = \log k_O + \left(\alpha + \frac{\gamma}{\sqrt{I}}\right)pH + \frac{\gamma}{\sqrt{I}} PZZP \tag{84}$$

where $\gamma \simeq 0.04$ is a constant, I is the ionic strength, and PZZP the point of zero zeta potential of the particle. Equation 84 shows that for low ionic strength the slope of the d log k_{ct}/dpH plot is increased and does not correspond straightforwardly to the transfer coefficient α.

According to Equation 76, the rate constant for a charge transfer between a semiconductor particle and a species in solution is size dependent. Frequently in a colloidal dispersion there is a distribution of particle sizes. Semiconductor particles having a different radius should react with different rate constants. The larger particles should react at a faster rate than the smaller ones. Such effects have indeed been observed.[111] Under pseudo-first-order conditions, where an exponential rate law is expected, polydispersity of the colloid is manifest by nonlinearity of the semilogarithmic plots, positive deviations occuring at longer times. A good curve fitting was obtained by assuming a Gaussian distribution of particle sizes.

Reactions of valence band holes of colloidal semiconductor with donors such as halides or thiocyanate (SCN^-) have also been investigated by pulsed laser techniques.[2,29,112] The oxidation of these species follows the sequence

$$X^- \xrightarrow{h^+} X\cdot \xrightarrow{X^-} X_2^{\bar{}} \tag{85}$$

and results in the formation of $X_2^{\bar{}}$ radical ions which are readily monitored by the characteristic absorption spectra. Kinetic studies show that the hole transfer occurs within the laser pulse (see Figure 25), indicating that the average time required for charge transfer is less than 10 nsec. Apparently, the oxidation involves only X^- species that are adsorbed to the semiconductor particle. The efficiency of the process follows the sequence $Cl^- > Br^- > SCN^- \simeq I^-$ and hence is closely related to the redox potential of the $X^-/X\cdot$ couple. The yield of $X_2^{\bar{}}$ radicals decreases sharply with increasing pH, becoming negligibly small at pH > 2.5. This is attributed to competing reaction of h^+ with surface OH groups (Equation 64). At higher pH, the latter process is thermodynamically favored over halide oxidation. Apart from this, desorption of X^- from the particle surface occurs as the pH approaches the PZZP. At pH 1 the quantum yields[2] are in the range of 0.08 ($Cl_2^{\bar{}}$) to 0.8 ($I_2^{\bar{}}$). Essentially similar results have been reported by Henglein,[112] although lower quantum yields were obtained. The yields of $Cl_2^{\bar{}}$ and $Br_2^{\bar{}}$ are greatly improved when RuO_2 is deposited onto the TiO_2 particles. Presumably, the RuO_2 plays the role of a hole transfer catalyst which increases the rate of charge transfer from the valence band of the semiconductor to the halide ion adsorbed on the surface. RuO_2 is commercially employed as an electrocatalyst for chlorine generation from chloride solutions.[113] Enhancement of hole transfer reactions by RuO_2 has been observed also with semiconductors other than TiO_2, such as $CdS^{29,114}$ and WO_3.[115]

An interesting experiment with Na_2CO_3-containing solutions of colloidal TiO_2 has recently been described by Chandrasekaran and Thomas.[116] Oxidation of carbonate by valence band

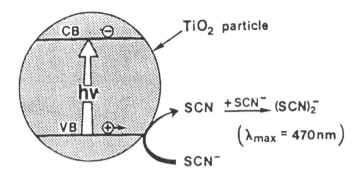

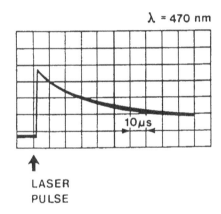

FIGURE 25.   Dynamics of interfacial hole transfer from the valence band of ultrafine TiO_2 particles to halide or pseudohalide ions adsorbed at the surface of the semiconductor. As an example, the hole reaction with thiocyanate is studied by exciting the TiO_2 colloid with the 347.1-nm laser flash. The oscillogram, recorded at 470 nm, shows that the formation of the oxidation product $(SCN)_2^-$ is immediate on the 10-μsec time scale employed in the experiment. The subsequent decay of the $(SCN)_2^-$ radicals occurs via disproportionation and back reaction with conduction band electrons.

holes and formation of CO_3^- radicals was analyzed by laser photolysis, monitoring characteristic absorption at 600 nm. Under steady-state UV light illumination formaldehyde was produced.

VI. PHOTOSENSITIZED ELECTRON INJECTION IN COLLOIDAL SEMICONDUCTORS

The photosensitization of electron transfer across the semiconductor solution interface plays a vital role in silver halide photography[117] and electroreprography.[118] Recently, it has also gained interest with regard to information storage and light energy conversion in photoelectrochemical cells.[119] In these systems, the phenomenon of photosensitized electron injection is used to effect charge separation with light of less than band gap energy. Of particular importance for the development of artificial photosynthetic devices is the sensitization of semiconducting oxides such as TiO_2,[120] SnO_2,[121] and ZnO_2.[122]

While much pertinent information has been gathered over the years on the overall performance of dye-sensitized semiconductor systems,[123] more precise data about the details of the electron injection process are urgently required. The rapid nature of these reactions requires application of fast kinetic techniques which, in the case of solid semiconductor electrodes or powders, is very difficult. On the other hand, the dynamics of electron injection can readily be investigated with colloidal semiconductors by application of flash photolysis. This is a relative new field, and only a few studies had appeared until recently.[123-126] In the following, we discuss the salient kinetic features of interfacial electron-transfer processes associated with sensitization. A system that has been thoroughly investigated consists of colloidal TiO_2 as a semiconductor and eosin (EO) as a sensitizer.[127] This is used as an example to illustrate the dynamics of the sensitization process.

Sensitized electron injection from excited eosin in the conduction band of TiO_2 occurs only under acidic condition (pH < 7), where the chromophore is strongly adsorbed at the semiconductor surface. The reactive excited state involved in the charge transfer reaction is the singlet and not the triplet state.[127] A similar result has been obtained by Kamat and Fox[125] for erythrosine sensitization of TiO_2. Due to oxidative quenching, the emission intensity of $EO(S_1)$ is strongly decreased in the presence of colloidal TiO_2. Laser excitation of EO was performed to identify the products resulting from the quenching reaction. Semioxidized EO, EO^+, was formed via electron injection from the excited singlet into the TiO_2 conduction band:

$$EO(S_1) \rightarrow EO^+ + e_{cb}^-(TiO_2) \tag{86}$$

This mechanism was confirmed by Rossetti and Brus[126] using time-resolved Raman spectroscopy. The quantum yield for charge injection was found to increase with decreasing pH. This effect can be understood by thermodynamic arguments: the driving force for reaction (3.86) is given by the difference between the redox potential of $EO(S_1)$ and the conduction band position of colloidal TiO_2 (Figure 26). For the former, a value of -1.2 V (NHE) is derived from the ground-state potential and the singlet excitation energy, while the latter is given by Equation 39. Thus, there is sufficient driving force for reaction 86, even in alkaline solution. Nevertheless, charge injection is only observed at pH 6, i.e., under conditions where EO is associated with the TiO_2 particles. This is reasonable, since close contact of the reactants is required for electron transfer to compete with the other channels of singlet deactivation, i.e., intersystem crossing, as well as radiative and nonradiative deactivation.

The quantum yield for charge injection was found to decrease at high occupancy of the TiO_2 particles by EO (average number of EO per particle > 30). This effect arises from concentration quenching which occurs under conditions where the distance of adjacent fluorophores approaches the critical radius for dipolar (Förster-type) energy transfer.

We have recently been able to follow the time course of charge injection directly by applying picosecond time resolved laser flash spectroscopy.[127] The observed rate for EO^+ formation (k_{obs}) is related to the rate constant for charge injection (k_{inj}) via Equation 87:

$$k_{inj} = k_{obs} \cdot \phi(EO^+) \tag{87}$$

where $\phi(EO^+)$ is the quantum yield for EO^+ formation. At pH 3, $k_{obs} = 2 \times 10^9$ sec^{-1}, and $\phi = 0.38$. The rate constant for charge injection is therefore $k_{inj} = 8 \times 10^8$ sec^{-1}. This is a very high rate for a heterogeneous charge transfer reaction, indicating strong exchange coupling between the excited singlet state of the EO and the d band states of the TiO_2 conduction band. The rate constant is only a factor of 60 below the limiting value of 5×10^{10} sec^{-1} predicted by Equation 83 for interfacial electron transfer over, distance of 5 Å at optimum driving force ($\Delta G = -\lambda$). At pH 3, Reaction 87 is thermodynamically

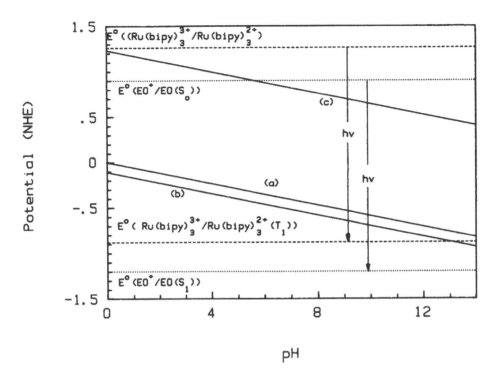

FIGURE 26. Redox potentials of ground states and excited states for EO (····) and Ru(bipy)$_3^{2+}$ (- - - -) in relation to the reversible potential of (a) the normal hydrogen electrode, (b) the conduction band potential of our colloidal TiO$_2$ particles, and (c) the normal oxygen electrode. This diagram illustrates that at pH 4 there is significant driving force for charge injection from the excited states of both dyes into the conduction band of TiO$_2$. The injected electron can produce hydrogen from water. However, only Ru(bipy)$_3^{2+}$ is capable of oxidizing water to O$_2$ at this pH.

downhill by about 1 eV (Figure 26). Thus, a relatively high expenditure of free energy is necessary to make the rate of electron injection competitive with other deactivation processes of the excited singlet state.

Apart from electron injection, the EO-colloidal TiO$_2$ system lends itself to a study of other important kinetic features of photosensitization of semiconductor particulates. Thus, it is possible to obtain information about light-induced charge separation in these systems. The question of how long the photoinjected conduction band electron can survive before it is recaptured by the dye cation radical is a primordial one for energy conversion. Again, laser photolysis investigations with colloidal semiconductor solutions give important clues as to the nature of these charge recombination processes.

Figure 27 shows oscillograms from flash excitation of aqueous EO solutions containing colloidal TiO$_2$. The TiO$_2$ concentration was chosen deliberately high in order to obtain an average occupancy of only one dye molecule per particle. The vertical rise in the absorbance at 470 nm reflects the formation of EO$^+$ during the laser pulse. The subsequent EO$^+$ decay is biphasic: a fast initial component is completed within about 10 μsec and is followed by a much slower decrease occurring over several hundred microseconds. The temporal behavior of the EO ground-state bleaching was also examined and was found to correspond to the mirror image of the EO absorption decay.

In order to rationalize these observations, we use the scheme shown in Figure 28A. Charge injection in the semiconductor is followed by electron recapture:

$$EO^+ + e_{cb}^-(TiO_2) \xrightarrow{k_b} EO \qquad (88)$$

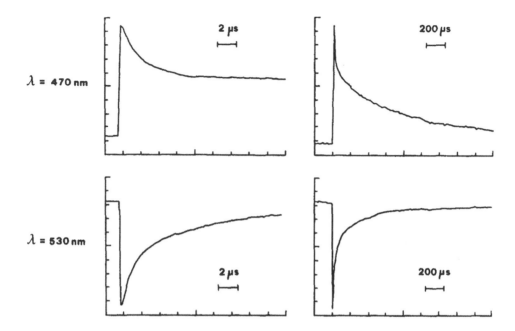

FIGURE 27. First direct observation of electron recapture from the conduction band of an ultrafine semiconductor by the parent dye cation absorbed onto the particle surface. Oscilloscope traces from the 532-nm laser photolysis of 10^{-5} M EO in aqueous colloidal TiO_2 (3g/ℓ); pH 3; no protective polymer was used. The temporal characteristics of the EO^+ decay at 470 nm and the ground-state bleaching at 530 nm is displayed on two different time scales. Note that the EO triplet absorbance makes a contribution to the 530 nm signal. (From Moser, J. and Grätzel, M., *J. Am. Chem. Soc*, 106, 6557, 1984. With permission.)

which was found to occur in two steps. The first part involves e_{cb}^- -EO pairs associated with the original host particle. Such an intraparticle process, in analogy to the intramicellar reactions discussed in Chapter 2, is expected to obey first-order kinetics if only one pair per particle participates in the reaction. In Figure 27 only about 50% of the EO cation radicals produced by the laser pulse undergo this type of electron recapture. The remaining fraction recombines on a much slower time scale, and this behavior is attributed to the desorption of EO^+ from the TiO_2 surface and subsequent bulk recombination. A detailed kinetic evaluation which takes into account the statistics of EO^+ distribution over the particles gives for rate constants of electron recapture and EO^+ desorption the values $k_b = 1.5 \times 10^5$ and $k_d = 3 \times 10^5$ sec^{-1}, respectively.

A comparison of the values obtained for k_b and k_d shows that for EO-TiO_2, electron injection occurs about 5000 times faster than charge recombination. This enables light-induced charge separation to be sustained on a colloidal TiO_2 particle for several microseconds.

Similar effects have been observed very recently with a number of different sensitizers, such as porphyrins[128] and Ru(bipy)$_3^{2+}$ derivatives[129] or Fe(CN)$_6^{4-}$ surface-derivatized TiO_2 particles.[130] In all these cases the back electron transfer from the semiconductor particle to the oxidized sensitizer occurred with a rate constant of 1 to 5 $\times$ 10^5 sec^{-1}, and it was several orders of magnitude slower than the forward injection. Thus, the combination of a sensitizer with a colloidal semiconductor particle affords a molecular device for light-induced charge separation. The question should be addressed why in this system the rate of the back reaction is so much smaller than that of the forward electron transfer. One might evoke the fact that recapture of the conduction band electron has a large driving force placing this reaction in the inverted region where the rate drops with increasing exothermicity. In addition, entropic factors should be considered. In a 120-Å-sized TiO_2 particle, the electron is delocalized over 3600 conduction band states. If there is only one sensitizer cation available at

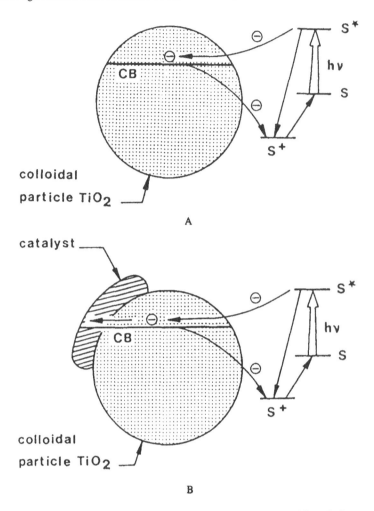

FIGURE 28. Schematic illustration of charge injection and intraparticle back electron transfer in the photosensitization of a colloidal semiconductor particle. (A) Without redox catalyst and (B) particles loaded with noble metal catalyst.

the surface, the back reaction is associated with a significant entropy decrease, which amounts to about 16 cal/°K mol.

Electrons injected in a colloidal semiconductor can be trapped by a noble metal deposit, such as Pt. This process is illustrated schematically in Figure 28B. In the presence of water, the electron trapped on a Pt site is rapidly converted into a hydrogen atom. Therefore, the noble metal deposit allows to suppress the fast intraparticle back reaction. These experiments offer the possibility to determine the rate of electron trapping and yield direct information on the nature of the junction (Schottky or ohmic) between the TiO_2 support and the noble metal deposit.

These investigations have meanwhile been extended to other sensitizers.[131-134] A very interesting case of potential practical importance is that of ruthenium tris (2,2'-bipyridyl-4,4'-dicarboxylate), $Ru[bipy(COO^-)_2]_3^{4-}$. By virtue of the carboxylate groups, this sensitizer is strongly adsorbed on the surface of TiO_2 in a pH domain between 2 and 5. Laser studies with colloidal dispersions allow the determination of the injection rate constant as $3.7 \times 10^7 \ sec^{-1}$. The quantum yield of charge injection is as high as 60 to 70%.[129] Electron recapture from the conduction band by the oxidized dye is much slower and occurs about at the same rate as in the case of EO^+. This Ru complex has a much higher oxidation potential than EO and hence is a candidate for a water cleavage sensitizer.

Table 5
PHOTOCATALYTIC AND PHOTOSYNTHETIC REACTIONS ON SEMICONDUCTOR PARTICULATE SYSTEMS

Studies with Naked Semiconductor Dispersions

Photoadsorption and photodesorption of gases
Photocatalytic oxidation of CO, H_2, N_2H_4, and NH_3
Isotopic exchanges on semiconductor surfaces
Photoproduction and decomposition of H_2O_2
Photooxidation and photoreduction of inorganic substances (CN^-, $S_2O_8^{2-}$, $Cr_2O_7^{2-}$, SO_3^{2-}...)
Photodeposition of metals
Photoreduction of CO_2 and of N_2
Photooxidation of organic materials (alkanes, alkenes, alcohols, aromatics...)
Photohydrogenation, photodehydrogenation reactions

Studies with Catalyst-Loaded Semiconductor Dispersions

Metallized semiconductor studies
  Photodecomposition of water
  Photooxidation of halides and cyanides
  Photooxidation of carboneous materials
  Photo-Kolbe reaction
  Photosynthetic production of amino acids
  Photochemical slurry electrode cells
  Photoassisted water-gas shift reaction
Metal-oxide coated semiconductor studies
  Photodecomposition of water
  Photodecomposition of H_2S

Table 6
APPLICATIONS

Catalysis	Paint and pigments	Optoelectronics
Catalyst supports O_2 activation and transfer	TiO_2 CdS Fe_2O_3	Q particles with low effective electron mass and high yield of luminescence
Photocatalysis Energy conversion (water splitting)	Removal of pollutants and toxic agents	Chalking of paints

VII. APPLICATIONS: PHOTOCATALYSIS

Semiconductor particulate systems have found numerous applications. In Table 5 we give some examples from the fields of catalysis, paints and pigments, and electronics. Photocatalysis is of particular interest in the context of the present monograph and will be discussed in a more detailed manner now.

There is currently extensive research activity in the area of heterogeneous photocatalysis with semiconductor particulate systems. Both naked and catalyst-loaded semiconductor dispersions are being assayed for various chemical processes. A distinction is sometimes made between photocatalytic ($\Delta G < 0$) and photosynthetic ($\Delta G > 0$) reactions. Table 6 lists some of these topics. Recent comprehensive reviews of these topics are available.[30,135-138] Here, we briefly present these areas which are of considerable practical interest and currently under investigation in our laboratory.

PARTICLE

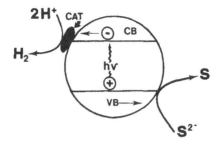

ELECTRODE

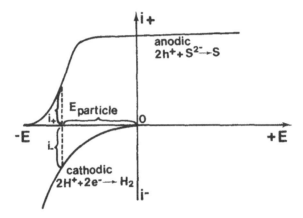

FIGURE 29. Operation of a semiconductor particle loaded with a catalyst as a light harvesting unit in the visible light-induced decomposition of H$_2$S.

A. Photocleavage of Hydrogen Sulfide

The cleavage of H$_2$S by visible light:

$$H_2S \rightarrow H_2 + S \qquad \Delta G^\circ = 8 \text{ kcal/mol} \tag{89}$$

is of industrial importance since H$_2$S occurs widely in natural gas fields and is produced in large quantities as an undesirable byproduct in the coal and petroleum industry.[139] It is a thermodynamically uphill reaction, and we illustrate in the following how this process can be driven by visible light using a semiconductor particle as a light harvesting unit. The H$_2$S cleavage process might be used in industrial procedures where H$_2$S or sulfides are formed as a waste product whose rapid removal and conversion into a fuel, i.e., hydrogen, are desired. Also, in an intriguing fashion, these systems mimic the function of photosynthetic bacteria that frequently use sulfides as electron donors for the reduction of water to hydrogen.

The operation of the semiconductor particle is illustrated schematically in Figure 29. Band gap excitation produces electron-hole pairs which diffuse to the surface of the particle where two redox reactions occur simultaneously. The valence band holes oxidize sulfide to sulfur, while the conduction band electrons reduce protons to hydrogen. In order to accelerate the latter reaction, one deposits a catalyst such as Pt, Rh, or RuO$_2$ on the particle surface. When sulfides such as CdS are used as semiconductors, the hole reaction with S^{2-} is very fast, and the overall process occurs with a very high quantum yield approaching 50%.[140-142]

In Figure 29 we have decomposed the overall reaction (Equation 89) in the two individual redox processes which take place simultaneously at the semiconductor surface. These can be analyzed separately by illuminating a n-type semiconductor electrode, made from the same material as the particle, in the sulfide solution. A photocurrent is observed at a voltage positive of the flat band potential, V_{fb}, when the electrode is polarized anodically. The cathodic dark current is determined with the semiconductor electrode onto which the same catalyst is deposited as on the particles. Since the overall current in the particle suspension is zero, we can use the cathodic and anodic current potential curves in the diagram to evaluate the particle potential under illumination. It corresponds to the potential for which i = 0 or $i_+ = i_-$. Furthermore, the reaction rate v can be derived from these current values via

$$v = -\frac{dn(H_2S)}{dt} = 2FSi_+ = 2FSi_- \qquad (90)$$

where F is Faraday's constant and S is the total surface of the particles present in solution.

The mechanism of photocatalysis by a semiconductor particle is reminiscent of that encountered in the catalysis of thermal redox reactions by colloidal metal particles. Consider, for example, the reduction of water to hydrogen by an electron relay R^- in the presence of a very small Pt particle (Figure 30). The role of the metal particle is to couple the oxidation of the relay to the reduction of water. Thus, an anodic and a cathodic reaction takes place simultaneously on the same particle. The electrochemical model of Wagner and Traud,[143] providing a quantitative description for local elements and mixed potentials in corrosion, is adequate in most cases to describe redox catalysis by colloidal metals.[144] The shape of the anodic and cathodic potential curve is determined by the overvoltage characteristics of the two couples on the conducting material acting as a catalyst. In the absence of diffusional effects, the Butler-Volmer equation applies (see Chapter 1, Section IV.C). The particle assumes a mixed potential E_p which corresponds to the intersection of the anodic and cathodic branch. At this potential, the overall reaction proceeds at a rate corresponding to the reaction current i_R.

This model gives a satisfactory description for many redox catalytic reactions, including the catalysis of the reduction of water to hydrogen by methylviologen radicals in the presence of colloidal Pt[145] and of the oxidation of water to oxygen in the presence of colloidal RuO_2.[146]

B. Photodegradation of Organic Wastes and Pollutants

Photocatalytic processes for the nonbiological conversion of materials which are common waste products of biomass and the biomass-processing industry to a fuel and/or chemicals are of great potential interest. Thus, photocatalytic reaction of glucose, the most common biomass building block, in the presence of TiO_2-RuO-Pt[147] or TiO_2-Pt[148,149] gives H_2 and CO_2. The mechanism of the process probably involves dehydrogenation of the hydroxyl functional groups of the organic compounds, and probably of aldehydes, with the formation of H_2 and carboxylic acids. These last compounds are then available for decarboxylation via a photo-Kolbe reaction resulting in CO_2 evolution and a shorter-chain alcohol which can continue to undergo oxidation. The end products are then only H_2, and CO_2.

A large variety of other biomass sources, such as proteins and fats (organic products in general), have also been examined.[148] The photocatalytic conversion into H_2, CO_2, and other various simple compounds has much potential in cleaning up stream wastes. Another class of compounds which is present in the environment as undesired products are the organochlorinated compounds.[149] Until now, little attention has been devoted to heterogeneous photoassisted catalytic degradation of such compounds,[150] and recently some haloaliphatic compounds were investigated.[151]

Haloaromatic compounds include several toxic and harmful materials. 4-Chlorophenol

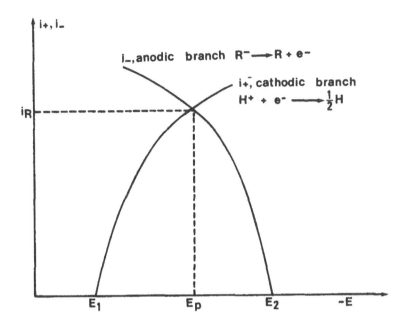

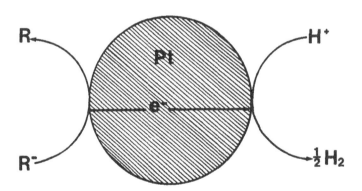

FIGURE 30.   Electrochemical description of redox catalysis by a small metal particle. The process considered is the reduction of water to hydrogen by a reduced electron relay R^-. E_1 and E_2 are the equilibrium redox potentials for the H^+-H_2 and R-R^- couples, respectively. E_p is the mixed potential assumed by the particle during the reaction and i_R the local current which defines the reaction rate.

degradation through a heterogeneous photoassisted process using TiO_2 suspensions has proved to give a complete mineralization into CO_2 and HCl.[152] The mass balance also precludes formation of other chlorine-containing organic or inorganic derivatives. In addition to light and the semiconductor, the presence of O_2 and water was essential for the process to proceed at a remarkable rate. The mechanism is likely to proceed through an oxygen-containing radical attack and concomitant Cl reduction to Cl^- by conduction band electrons. Further oxidation of the dihydroxybenzene ring breakdown leads to the final simple products. It is noteworthy that a solar experiment, performed on a sunny day in Torino, Italy (70 to 75 mW/cm²), showed a complete degradation of 4-chlorophenol within a few hours. It is worthwhile to remember that, since the extent of toxicity is related to the chlorine content, even partial dechlorination would prove useful for the degradation of environmentally dangerous chlorinated pollutants.

Noteworthy progress has recently been achieved in the photocatalytic destruction of phosphate-esters used as warfare agents and insecticides.[153]   For example, diethyl nitrophenyl phosphate is destroyed within minutes on the surface of TiO_2 (anatase) particles under illumination with a solar simulator. By contrast, when the same compound is deposited onto SiO_2 particles, no degradation occurs even within 50 hr of illumination. These findings illustrate the high activity of the TiO_2 photocatalyst in promoting the destruction of the phosphate ester.

Finally, since photoprocesses involving inorganic pollutants such as SO_3^{2-}, CN^-, or heavy metal ions have been already reported and photodegradation of surfactants[154] recently achieved, the prospects for potential application of these processes are indeed bright.

C. Photodecomposition of Water: Solar Energy Conversion

A process which is of particular importance for future energy supplies is the photodecomposition of water into hydrogen and oxygen:

$$H_2O \xrightarrow{h\nu} H_2 + 0.5O_2 \qquad \Delta G^\circ_{298} = 237 \text{ kJ/mol} \tag{91}$$

Hydrogen is an energy vector with very desirable properties. First, it is a powerful fuel: the energy storage capacity is 120,000 J/g as compared to 40,000 for oil and 30,000 for coal. Furthermore, hydrogen has the advantage over conventional (fossil) or nuclear energy sources in that combustion in a thermal engine or a fuel cell does not result in pollution of the environment. Reaction 91 requires an input of 237 kJ of free enthalpy for each mole of hydrogen produced under standard conditions.

The following economic analysis is instructive. Consider a solar collector containing a catalytic system that performs Reaction 91 with a 10% overall efficiency. At an average solar irradiance of 200 W/m², 66 m³ of hydrogen can be obtained per year from each square meter of collector surface. At the current price for hydrogen of $0.2/m³, this would represent a value of $13.00. If a 10% return on the initial investment per year is considered to be reasonable, the expenses for the installation of the water-splitting unit should be not higher than $130/m². The fulfillment of this condition could well be in reach by employing systems that are based on colloidal semiconductor dispersions or polycrystalline electrodes.

The efficiency of a solar energy converter is given by the ratio of work produced to incident power. The incident power is obtained by integrating over the solar emission for which the spectral distribution is given in Figure 31. Note that the solar emission spectrum depends on atmospheric conditions. In outer space, one observes the characteristics given by the curve labeled AM 0. Through scattering and absorption in the atmosphere, the spectrum is modified, and this is expressed by the air mass number defined by the relationship AM = 1/sin δ, where δ denotes the angle between the surface of the earth and the incident light beam. The pathway of sunlight through the atmosphere is shortest when δ = 90°, i.e., AM = 1, and Figure 31 shows the spectral distribution for this case.

In order to derive an expression for the efficiency of conversion of light into chemical energy, we first calculate the flux of solar photons that is absorbed by the light energy converter. The majority of endergonic photochemical reactions have a threshold wavelength, λ_g; photons of energy less than this value cannot be absorbed and hence cannot contribute to the energy conversion. If a semiconductor is used as a light harvesting unit, the relation between band gap energy, E_g, expressed in electron volts, and threshold wavelength, is λ_g (nm) = $1240/E_g$. Let $N(\lambda)$ be the incident solar flux in the wavelength region between λ and $\lambda + d\lambda$, and $\alpha(\lambda)$ the extinction coefficient of the absorber. The absorbed flux of photons is then given by

$$J_{abs} = \int_0^{\lambda_g} N(\lambda)\alpha(\lambda)\, d\lambda \tag{92}$$

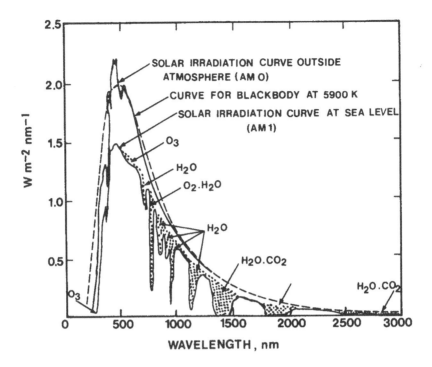

FIGURE 31.   The solar emission spectrum.

The available solar power E (W/m^2) at the band gap wavelength λ_g is given by

$$E = J_{abs} \frac{hc}{\lambda_g} \qquad (93)$$

where h is Planck's constant and c the velocity of light. The fraction η_E of incident solar power available to initiate photochemistry, then, is

$$\eta_E = \frac{E}{\displaystyle\int_0^\infty E(\lambda) \, d\lambda} \qquad (94)$$

where the denominator represents the total solar power. Curve I in Figure 32 is a plot of this η_E (for AM 1.2 radiation) as a function of λ_g, for the ideal case where $\alpha(\lambda) = 1.0$ for $\lambda \leq \lambda_g$. The η_E has a maximum value of $\sim 47\%$ at 1110 nm, but it is rather broad ($>45\%$ over the wavelength range of 800 to 1300 nm). Curve I represents an ideal limit; in real life, there are further thermodynamic and kinetic limitations on the conversion of light energy into chemical energy. These have been examined in detail by Ross and Calvin,[155]Bolton,[156] Almgren,[157] and others. Taking into account the thermodynamic limitations alone, in curve II in Figure 32, the maximum drops to about 32% at a threshold wavelength of 840 nm.

For conversion of sunlight to electricity, the thermodynamic limit may well be approached. For example, a gallium arsenide solar cell has been reported with an efficiency (AM, 1.4) of 23% ($\lambda_g = 920$ nm). However, if instead of electricity production one proposes to drive an endonergic chemical reaction such as the cleavage of water by light, the requirement of energy storage imposes additional constraints. As a result, the efficiency of solar energy conversion will be further reduced. These kinetic or overvoltage losses depend on the reaction rate as well as on the activity of the catalysts required to convert redox equivalents into

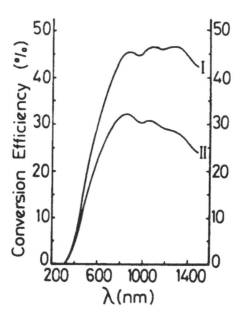

FIGURE 32. Variation in the conversion efficiency with wavelength for photochemical devices with a threshold absorption wavelength λg. Curve I is a plot of the fraction of incident solar power (%) that is available at various threshold wavelengths. Curve II is a plot of the thermodynamic conversion efficiencies under optimal rates of energy conversion (data for AM 1.2 radiation).

fuels. For example, in the water cleavage process, electrons are used to produce hydrogen from water, while the positive charge carriers perform the oxidation of water to oxygen. These reactions require the presence of a catalyst, and the overvoltage losses in these catalystic reactions are expected to lie between 0.4 and 0.8 V, depending on the rate of water decomposition, i.e., on light intensity. If 0.8 V per absorbed photon must be sacrificed for light energy storage, the threshold wavelength for carrying out the water photolysis is 611 nm and the maximum overall efficiency is about 17%. Considerably higher efficiencies can be achieved with devices where two photosystems operate in series, e.g., natural photosynthesis or tandem photovoltaic cells.[156]

Three suitable microheterogeneous light harvesting units capturing photons, converting energy into chemical potential, and using it to decompose water are shown in Figure 33. Figure 33A represents a two-component system containing a sensitizer (S) and an electron relay (R). Light promotes electron transfer from S to R, whereby the energy-rich radical ions S^+ and R^- are produced. Thus, light functions here as an electron pump operating against a gradient of chemical potential. A similar process may be conducted in a system where the sensitizer acts as the electron acceptor in the excited state and the electron relay is oxidized. Such photoinduced redox reactions have been studied in great detail both from the experimental[158] and the theoretical[159] point of view. The rate of electron transfer between the excited sensitizer and relay is expected to approach the diffusion-controlled limit as soon as the driving force for the reaction exceeds a few hundred millivolts. Conversely, the backward electron transfer between S^+ and R^-, which is thermodynamically strongly favored, is almost always diffusion controlled. This poses a severe problem for the use of such systems in energy-conversion devices as it limits the lifetime of the radical ions to at

a/ SENSITIZER / RELAY PAIR

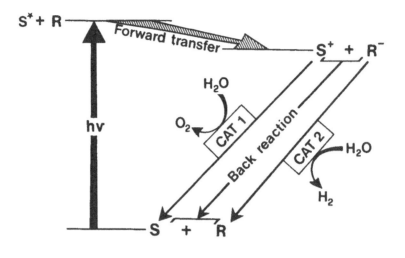

b/ SENSITIZER / COLLOIDAL SEMICONDUCTOR

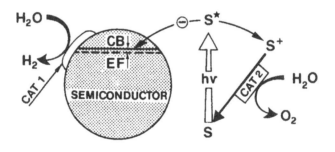

c/ COLLOIDAL SEMICONDUCTOR

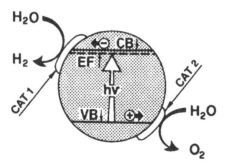

FIGURE 33.   Light harvesting units for water photolysis in microheterogeneous systems. (a) Sensitizer-relay pair; (b) sensitizer-colloidal semiconductor; and (c) colloidal semiconductor.

most several milliseconds under solar light intensity. In Chapter 2 we gave examples of how suitable molecular assemblies may be developed that allow the retardation of this undesired back reaction.

With regard to the choice of the sensitizer and the relay, compounds must be found that are suitable from the viewpoints of both light absorption and redox potentials, and that undergo no chemical side reactions in the oxidation states of interest. The sensitizer should have good absorption features with respect to the solar spectrum. Also, the excited state should be formed with high quantum yield and have a reasonably long lifetime, and the electron-transfer reaction must occur with high efficiency, i.e., good solvent cage escape yield of the redox products. The redox properties of the donor-acceptor relay must obviously be tuned to the fuel-producing transformation envisaged. If, for example, water cleavage by light is to be achieved, then the thermodynamic requirements are such that $E° (S^+/S) >$ 1.23 V (NHE) and $E° (R/R^-) < 0$ V under standard conditions. In the design of sensitizer-relay couples suitable for photoinduced water decomposition, considerable progress has been made over the last few years.[30] A number of systems converting more than 90% of the threshold light energy required for excitation of the sensitizer into chemical potential have been explored. Also, in several cases the reduced relay and oxidized sensitizer are thermodynamically capable of generating H_2 and O_2 from water:

$$R^- + H_2O \rightarrow 0.5H_2 + OH^- + R \tag{95}$$

$$2S^+ + H_2O \rightarrow 0.5O_2 + 2H^+ + 2S \tag{96}$$

Noteworthy examples are sensitizers such as $Ru(bipy)_3^{2+}$ and derivatives, while viologens, metal ions, or metal ion complexes have been used as relays. The catalysts employed are usually ultrafine particles: RuO_2 for water oxidation and, typically, Pt for water reduction. There are significant kinetic problems with the system in Figure 33A, since the intervention of these catalysts must be rapid and specific at the same time. Furthermore, the cross reaction of the reduced relay with O_2 will lead to a photostationary state where the water photolysis is short-circuited. A recent example for the successful embodiment of such a system involves RuO_2-loaded sepiolite as support, Eu^{3+}-substituted Al_2O_3 particles as the relay, and Pt as the hydrogen evolution catalyst.[160]

A second type of light harvesting unit suitable for water cleavage by visible light is illustrated in Figure 33B. The sensitizer is adsorbed onto a colloidal semiconductor particle, and no electron relay is required. The excited state of the sensitizer injects an electron into the conduction band of the semiconductor, where it is channeled to a catalytic site for hydrogen evolution. A second catalyst mediates oxygen generation from S^+ and H_2O, whereby the original form of the sensitizer is regenerated. Several successful attempts to split water by visible light in this way have recently been described,[161] and earlier work has been reviewed.[30] These systems offer promising prospects, in particular in view of recent developments[162] which include the use of a dimeric ruthenium-EDTA complex as a water oxidation catalyst.[163]

In the third configuration, shown in Figure 33C, the semiconductor particle is excited by band gap illumination, forming electron-hole pairs. The former afford hydrogen formation, while the latter give rise to O_2 generation. There are many examples for the photodecomposition of water with this type of system.[30] In most cases, TiO_2 or $SrTiO_3$ were used as a semiconductor material, and this restricts the photoactivity to the UV region of the spectrum. However, lower band gap materials such as CdS[164] and V_2O_5[165] have recently been employed and show activity in the visible. In this case, a highly active water oxidation catalyst is required to promote water oxidation at the expense of corrosion.

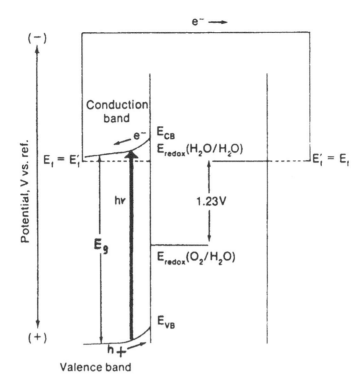

FIGURE 34.   Photoelectrolysis of water by using an n-type semiconductor electrode.

Apart from semiconductor particles, it is possible to cleave water by sunlight using semiconductor electrodes or membranes. The principle of such a device is shown in Figure 34, where we have chosen the case of an n-type semiconductor electrode as an example. When such an electrode is brought in contact with an aqueous electrolyte, negative charge carriers are transferred spontaneously from the semiconductor to the solution phase, resulting in the formation of a Schottky-type potential barrier. As a consequence, an electrical space charge, i.e., a depletion layer, is built up underneath the surface of the semiconductor. Light is absorbed mostly in this region, and the presence of an electrical field separated the electrons from the holes (the situation is analogous to that of the p-n junction in a photovoltaic cell). Holes are driven to the surface where oxidation of water to oxygen occurs, while the electrons migrate through the back contact of the electrode and the external circuit to the counterelectrode, where hydrogen is evolved. The energetic position of conduction and valence bands must be such that they straddle the water reduction, $E_r(H_2O/H_2)$, and oxidation, $E_r(H_2O/O_2)$, potentials. Only a few semiconducting materials, such as TiO_2 (anatase), CdS, and $SrTiO_3$, satisfy this condition. An additional requirement is that the semiconductor does not decompose under illumination. While this is the case for TiO_2[166] and $SrTiO_3$, the band gap of these materials is too large to allow for efficient solar energy conversion.

The response of wide oxide semiconductor photoelectrodes to solar light can be enhanced dramatically by chemisorbed dyes. The salient features of spectral sensitization were discussed in the previous paragraph in connection with colloidal semiconductors. The operation of a sensitized photoelectrochemical device containing an n-type semiconductor is outlined in Figure 35. The dye must be selected such that the ground-state redox potential lies within the band gap of the semiconductor, while that of the excited state lies above the conduction band edge.[167,168] It is clear from the diagram that a photon energy less than the band gap of the semiconductor may still excite the dye molecule to this state, from which a possible decay mode is by electron loss to the n-type semiconductor. The dye molecule, now in an

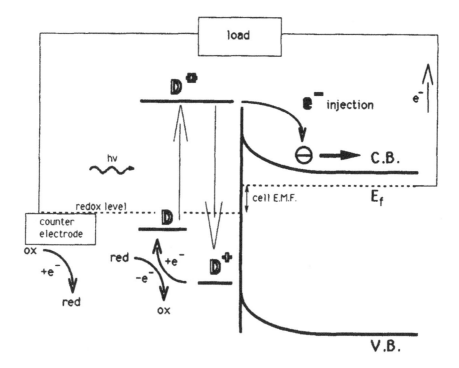

FIGURE 35. Mechanism of dye sensitization and charge transfer in a sensitized regenerative photoelectrochemical cell.

oxidized state, returns to the initial state by reaction with an electron donor (reducing) agent present in the electrolyte. At the cathode, a reduction reaction takes place. If the anode reaction is the reverse of the cathodic one, then the cell is a current producing system, analogous to a solid-state photovoltaic cell. If this is not the case, for example, if the two reactions are, respectively, the oxidation and reduction of water, then the cell is photosynthetic and produces oxygen and hydrogen gases.

A system which meets these energy level conditions may nonetheless remain inefficient on mechanistic grounds. For example, if the dye is merely dissolved in the electrolyte, the excited state is so rapidly quenched, typically within 10^{-8} sec, that diffusion to the electrode surface and charge transfer may be ruled out.[168] Surface-attached species alone can contribute to the sub-band gap photoresponse of the device. On such a modified semiconductor surface, only the first adsorbed monolayer can transfer charge, thicker dye layers tending to be insulating. However, on a flat surface optical absorption by monolayers is weak, at most a few percent, so that the sensitization effect is feeble. We have developed a system where the TiO_2 surface is highly porous (roughness factor, about 200), and the dye is chemically modified to enhance adsorption. Below, we show representative dye structures employed in these studies.[128-130]

In chromophores A and C, the adhesion to the TiO_2 surface is greatly enhanced by the carboxylate groups. In B, the Ru complex is chemically attached via oxygen bridges. Ferrocyanide (D) forms a charge transfer complex with Ti^{IV} ions at the TiO_2 surface, which greatly enhances the visible light response of this semiconductor.[130]

On the porous TiO_2 layer (the fractal dimension of the layer is approximately 2.7), there is in consequence sufficient dye present for effective absorption of visible light, while retaining intimate contact with the semiconductor to facilitate charge transfer. Transport losses in the semiconductor are minimal. Though functionally similar to a conventional n-type photoanode, the dye-sensitized semiconductor operates by electron injection and is therefore a majority carrier device. The high recombination losses due to a disorder in a semiconductor structure where the photoexcited electroactive charge carriers are holes are not encountered in the present case. The effect was known in principle following the work of Matsumura et al.[169] and Alonso et al.[170] on sintered ZnO electrodes sensitized by rose bengal dye, and was recently developed by us[171] to display an unprecedented incident photon to current conversion efficiency of up to 60% in the visible.

As an example, we show in Figure 36 the action spectrum for the photocurrent observed during visible light irradiation of $Fe(CN)_6^{4-}$-coated TiO_2 electrodes.[130] The solution contained hydroquinone as the electron donor. The action spectrum is characterized by a maximum at 420 nm, matching the absorption features of the $Fe(CN)_6^{4-}$ complex formed with colloidal TiO_2 particles. The incident monochromatic photon-to-current conversion efficiency, defined as the number of electrons injected by the excited sensitizer (and recorded as photocurrent) divided by the number of incident photons was calculated from the equation

$$\eta\ (\%) = \frac{1.24 \times 10^3 \times \text{photocurrent density } (\mu A/cm^2)}{\text{wavelength (nm)} \times \text{photon flux } (W/m^2)}$$

The current density obtained at 420 nm at an incident light flux of 0.9 W/m^2 is 10.4 $\mu A/cm^2$, which corresponds to $\eta = 37\%$. In the absence of $Fe(CN)_6^{4-}$ $\eta = \leq 2\%$ for $\lambda > 400$ nm. This shows the remarkable efficiency of $Fe(CN)_6^{4-}$ for sensitizing the visible light response of TiO_2. A monolayer of $Fe(CN)_6^{4-}$ deposited on a flat electrode would absorb less than 1% of the incident light. The high roughness factor of the TiO_2 layer enhances the harvesting of visible photons by allowing the sensitizer to attain a several hundred times larger concentration than on a smooth surface. Multiple reflection of the light within the fractal structure of the layer could also contribute to the high efficiency.

The insert in Figure 36 presents the photocurrent potential curve measured during illumination (440-nm light) of the $Fe(CN)_6^{4-}$-coated electrode under conditions similar to those of the photocurrent action spectrum. The steady-state photocurrent has the onset at -0.2 V (SCE), which is close to the flat band potential of TiO_2. It rises steeply with increasing potential, reaching a plateau at 0.1 V. This indicates that high fill factors can be obtained

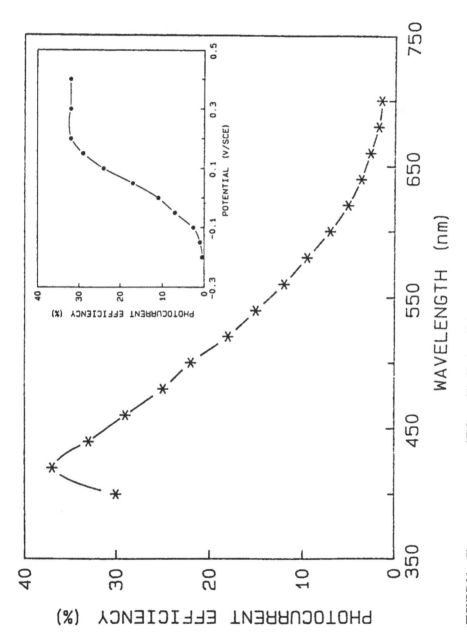

FIGURE 36. Photoresponse spectrum of TiO₂ sensitized by iron (II) hexacyanide complex. The insert shows dependence of photocurrent on applied potential. All photoresponse spectra are measured in the plateau (anodic bias) region.

with this type of electrode in the conversion of light to electricity. Preliminary tests with a cell configuration such as shown in Figure 35 indicate power conversion efficiencies exceeding 12% in the visible.

Photoelectrochemical water cleavage can also be performed with p-type semiconductors. High efficiencies ($\sim$ 12%) have been obtained by Heller [172] with p-InP single-crystal electrodes covered with a noble metal catalyst for water reduction to hydrogen.

D. Incorporation of Colloidal Semiconductors in Supramolecular Assemblies

Fascinating work is presently being carried out on the use of colloidal semiconductors as building blocks for supramolecular assemblies. These studies are part of the rapidly developing field[173] of molecular electronics.

Topics under investigation include light-induced charge separation, the vectorial transfer of electrons in space, and the switching of transmembrane ion fluxes by photoexcitation of a membrane constituent. The use of colloidal semiconductors in these devices could be advantageous in that the ultrafine particle could play the role of the light harvesting unit or assist in the rapid displacement of electronic charge. The semiconductor could furthermore induce redox transformations of molecules, as discussed in the previous sections, or provoke structural changes such as *cis-trans* isomerizations.[174] The latter reaction is important in the permeability control of membranes.[175]

This field is still in infancy. Nonetheless, significant progress has already been achieved in combining semiconductors in supramolecular assemblies. For example, Tricot and Fendler [176] succeeded in incorporating colloidal semiconductors, e.g., CdS, in inverted micelles and vesicles. The oxidation of a donor such as thiophenol or benzylalcohol by photogenerated valence band holes was coupled to hydrogen generation by conduction band electrons using Rh as a redox catalyst. When the CdS particles were produced at the inner side of the vesicle wall, the reactivity was much larger than at the outside. The water pockets present in inverted micelles and vesicles can also be used to control the growth of semiconductor colloids during preparation. This was already discussed in connection with quantum size effects.[25,26]

Thomas[24] has investigated the properties of quasi-linear CdS formed in the ultrafine channels of porous Vicor glass. Kakuta et al.[177] have precipitated colloidal CdS onto Nafion membranes and investigated light-driven redox reactions such as the reduction of methylviologen and the cleavage of H_2S. These first results show the feasibility of combining the advantages of semiconductors and molecular assemblies and provide an important impetus for further investigations.

VIII. CONCLUDING REMARKS

The investigation of photochemical reactions in semiconductor dispersions has a long history which dates back to the invention of silver halide photography and the work on oxides by Eibner[178] and Tamman[179] at the beginning of this century. The discoveries of Baur and Perret in Switzerland[180] provided an important step toward the understanding of these processes. Thus, Baur and co-workers were the first to show that aqueous ZnO dispersions produced oxygen under illumination in the presence of electron acceptors such as Ag^+ ions. The results were correctly interpreted in terms of a molecular electrolysis model, i.e., the photogeneration of oxidizing and reducing equivalents and the subsequent reaction with water and Ag^+ ions, respectively.

More recently, this field has encountered a very rapid development, mainly due to the importance for light energy conversion and photocatalysis. Krasnovsky and Brin,[181] produced oxygen from water by visible light irradiation of aqueous WO_3 dispersions where Fe^{3+} ions served as electron acceptors. Very interesting discoveries in the field of photocatalysis of organic reactions by semiconductor particles were made by Bard.[137,138] The studies of

Gerischer[182] and Fujishima and Honda,[166] performed with semiconducting electrodes instead of particles, provided the impetus for the development of photoelectrochemistry, a new field which counts more than 1300 publications over the last 8 years.[183] To cover this domain would have gone much beyond the scope of the present monograph.

A wealth of knowledge has been acquired in the domain of photochemical heterogeneous electron-transfer processes during the last 10 years. Research in this area continues to advance at a very rapid pace, a recent highlight being the incorporation of colloidal semiconductors in molecular assemblies such as vesicles and inverted micelles and the efficient light energy conversion on fractal TiO_2 layers. The exploration of organized molecular assemblies as minute reaction media to kinetically control the dynamics of photoinitiated redox events is another focal point in the field of artificial photosynthesis.

The discovery of highly active redox catalysts allowed for the coupling of light-induced charge separation to fuel-generating steps. Following the initial work on ultrafine Pt and RuO_2 particles as mediators for the reduction and oxidation of water, respectively, this field has traversed a very active phase. The essential factors controlling the activity of the noble metal particles are established, and a significant effort has gone into optimizing these systems, perhaps apart from the oxygen generation catalyst, which still needs to be improved.

The advent of colloidal semiconductor particles has rendered feasible the direct observation of light-induced charge carrier generation, the recombination and reaction with catalysts or molecular acceptors. These investigations will continue to thrive, and more sophisticated assemblies, such as surface-derivatized particles, will soon be scrutinized. Water cleavage by visible light remains the primary target of such studies, although other processes, e.g., the photochemical splitting of H_2S, are also of great interest. Much remains to be done, and the task is perhaps one of the most challenging facing scientists.

REFERENCES

1. Hayashi, S., Nakamori, N., Kanamori, H., Yodogawa, Y., and Yamamoto, K., *Surf. Sci.*, 86, 665, 1979.
2. Moser, J. and Grätzel, M., *Helv. Chim. Acta*, 65, 1436, 1982.
3. Stäber, W., Fink, A., and Bohn, E., *J. Colloid Interface Sci.*, 26, 62, 1982.
4. Turkevich, J., Aika, K., Ban, L. L., Okura, I., and Namba, S., *J. Res. Inst. Catal. Hokkaido Univ.*, 24, 54, 1976.
5. Matijevic, E., *Langmuir*, 2, 12, 1986.
6. Dutton, D., *Phys. Rev.*, 112, 785, 1985.
7. Ramsden, J. J. and Grätzel, M., *J. Chem. Soc. Faraday Trans. 1*, 80, 919, 1984.
8. Ramsden, J. J., Webber, S. E., and Grätzel, M., *J. Phys. Chem.*, 89, 2740, 1985.
9. Kreibich, U. and Fragstein, C., *Z. Phys.*, 224, 307, 1969.
10. Papavassiliou, G. C., *Prog. Solid State Chem.*, 12, 185, 1979.
11. Hsu, W. P. and Matijevic, E., *Appl. Opt.*, 24, 1623, 1985.
12. Frölich, H., *Physica*, 6, 406, 1937.
13. Kubo, R., Kawabata, A., and Kabayashi, S., *Ann. Rev. Mater. Sci.*, 14, 49, 1984.
14. Bauer, G., Kucher, F., and Heinrich, H., Eds., *Two-Dimensional Systems: Heterostructures and Superlattices* (Springer Series in Solid State Sciences), Springer-Verlag, Berlin, 1984.
15. Taeckel, G. V., *Z. Tech. Phys.*, 7, 301, 1926.
16. Inmau, J. K., Mraz, A. M., and Weyl, W. A., *Solid Luminescent Materials*, John Wiley & Sons, New York, 1984, 182.
17. Berry, C. R., *Phys. Rev.*, 161, 848, 1967.
18. Stasenko, A. G., *Sov. Phys. Solid State*, 10, 186, 1968.
19. Skomyakov, L. G., Kitaev, G. A., Shcherbakova, Ya., and Belyaeva, N. N., *Opt. Spectrosc.*, 44, 82, 1978.
20. Nedeljkovic, J. M., Nenadovic, M. T., Micic, O. I., and Nozik, A. J., *J. Phys. Chem.*, 90, 12, 1986.
21. Foitik, A., Weller, H., and Henglein, A., *Chem. Phys. Lett.*, 120, 552, 1985.

22. Koch, U., Foitik, A., Weller, H., and Henglein, A., *Chem. Phys. Lett.*, 122, 507, 1985.
23. Weller, H., Foitik, A., and Henglein, A., *Chem. Phys. Lett.*, 117, 485, 1985.
24. Thomas, J. K., *J. Phys. Chem.*, 91, 207, 1987.
25. Watzke, H. J. and Fendler, J. H., *J. Phys. Chem.*, 91, 854, 1987.
26. Brus, L. E., *J. Chem. Phys.*, 80, 4403, 1984; *J. Chem. Phys.*, 79, 5566, 1983.
27. Foitik, A., Weller, H., Koch, U., and Henglein, A., *Ber. Bunsenges. Phys. Chem.*, 88, 969, 1984.
28. Weller, H., Schmidt, H. M., Koch, U., Foitik, A., Baral, S., Henglein, A., Kunath, W., Weiss, K., and Diekmann, E., *Chem. Phys. Lett.*, 124, 557, 1986.
29. Duonghong, D., Ramsden, J. J., and Grätzel, M., *J. Am. Chem. Soc.*, 104, 2977, 1982.
30. Kalyanasundaram, K., Grätzel, M., and Pelizzetti, E., *Coord. Chem. Rev.*, 69, 57, 1986.
31. Becker, W. G. and Bard, A. J., *J. Phys. Chem.*, 87, 4888, 1983.
32. Rossetti, R. and Brus, L. E., *J. Phys. Chem.*, 86, 4470, 1986.
33. Vuysteke, A. A. and Sihronen, Y. T., *Phys. Rev. C*, 113, 400, 1959.
34. Serpone, N., Sharma, D. K., Jamieson, M. A., Grätzel, M., and Ramsden, J. J., *Chem. Phys. Lett.*, 115, 473, 1985.
35. Rossetti, R. and Brus, L. E., *J. Phys. Chem.*, 90, 558, 1986.
36. Memming, R., *J. Electrochem. Soc.*, 116, 786, 1969.
37. Beckmann, K. and Memming, R., *J. Electrochem. Soc.*, 116, 368, 1969.
38. Petermann, G., Tributsch, H., and Bogomolni, R., *J. Chem. Phys.*, 57, 1026, 1972.
39. Ellis, A. B. and Kavas, B. R., *J. Am. Chem. Soc.*, 101, 236, 1979.
40. Kavas, B. R. and Ellis, A. B., *J. Am. Chem. Soc.*, 102, 968, 1980.
41. Van Ryswyk, H. and Ellis, A. B., *J. Am. Chem. Soc.*, 108, 2454, 1986.
42. Nakato, Y., Tsumura, A., and Tsubomura, H., *Chem. Phys. Lett.*, 85, 387, 1982.
43. Nakato, Y., Tsumura, A., and Tsubomura, H., *J. Phys. Chem.*, 87, 2402, 1983.
44. Nakato, Y., Ogawa, H., Morita, K., and Tsubomura, H., *J. Phys. Chem.*, 90, 6210, 1986.
45. Nakato, Y., Morita, K., and Tsubomura, H., *J. Phys. Lett.*, 90, 2718, 1986.
46. Papavassiliou, G. C., *J. Solid State Chem.*, 40, 330, 1981.
47. Chestnoy, N., Harris, T. D., Hull, R., and Brus, L. E., *J. Phys. Chem.*, 90, 3393, 1986.
48. Thomas, D. G., Hopfield, J. J., and Augustyniak, W. M., *Phys. Rev.*, 140, A202, 1965.
49. Henry, C. H., Faulkner, R. A., and Nassau, K., *Phys. Rev.*, 183, 798, 1969.
50. Era, K., Shionoya, S., Washizawa, Y., and Ohmatsu, H., *J. Phys. Chem. Solids*, 29, 1843, 1968.
51. Taguchi, T., Yokogawa, T., and Yamashita, H., *Solid State Commun.*, 49, 551, 1984.
52. Moroz, M., Brada, Y., and Honig, A., *Solid State Commun.*, 47, 115, 1983.
53. Broser, I., Gutowski, J., and Reidel, R., *Solid State Commun.*, 49, 445, 1984.
54. Jortner, J., *J. Chem. Phys.*, 64, 4860, 1976.
55. Jortner, J., *Biochem. Biophys. Acta*, 594, 193, 1980.
56. Gerischer, H., in *Photoelectrochemistry, Photocatalysis and Photoreactors* (NATO ASI Ser. C), Vol. 146, Schiavello, M., Ed., D. Reidel, Dordrecht, Netherlands, 1985.
57. Memming, R., in *Comprehensive Treatise in Electrochemistry*, Vol. 7, Conway, B. E. et al., Eds., Plenum Press, New York, 1983, 529.
58. Hunter, R. J., Ed., *Zeta Potential in Colloidal Science*, Academic Press, London, 1981.
59. Ward, M., White, J., and Bard, A. J., *J. Am. Chem. Soc.*, 105, 27, 1983.
60. Albery, W. J. and Bartlett, P. N., *J. Electrochem. Soc.*, 131, 315, 1984.
61. Gärtner, W., *Phys. Rev.*, 116, 84, 1959.
62. Curran, J. S. and Lamouche, D., *J. Phys. Chem.*, 87, 5405, 1983.
63. Mitchell, J. W., *Photo. Sci. Eng.*, 27, 3, 1983.
64. Fujishima, A., Inoue, T., and Honda, K., *J. Am. Chem. Soc.*, 101, 5582, 1979.
65. Ward, M. D. and Bard, A. J., *J. Phys. Chem.*, 86, 3599, 1982.
66. Fuji, M., Kawai, T., and Kawai, S., *Chem. Phys. Lett.*, 106, 517, 1984.
67. Albery, W. J., Bartlett, P. N., and Porter, J. D., *J. Electrochem. Soc.*, 131, 2892, 1984.
68. Perone, S. P., Richardson, J. H., Deutscher, S. B., Rosenthal, J., and Ziemer, J., *J. Electrochem. Soc.*, 127, 2580, 1980; *J. Phys. Chem.*, 85, 341, 1981.
69. Herzion, Z., Croitoru, N., and Gottesfeld, S., *J. Electrochem. Soc.*, 128, 551, 1981.
70. Gottesfeld, S. and Feldberg, S. W., *J. Electroanal. Chem. Interfacial Electrochem.*, 146, 47, 1963.
71. Kamat, P. V. and Fox, M. A., *J. Phys. Chem.*, 87, 59, 1983.
72. Prybyla, S., Struve, W. S., and Parkinson, B. A., *J. Electrochem. Soc.*, 131, 1587, 1984.
73. Itoh, K., Nakao, M., and Honda, K., *J. Appl. Phys.*, 57, 5493, 1985.
74. Jaegermann, W., Sakata, T., Janata, E., and Tributsch, H., *J. Electroanal. Chem. Interfacial Electrochem.*, 189, 65, 1985.
75. Wilson, R. H., Sakata, T., Kawai, T., and Hashimoto, K., *J. Electrochem. Soc.*, 132, 1082, 1985.
76. Friplat, A. and Kisch-De Mesmaeker, A., *J. Electrochem. Soc.*, 134, 66, 1987.
77. Bitterling, K. and Willig, F., *J. Electroanal. Chem.*, 204, 211, 1986.

78. Wilkingson, F. and Willsher, C. J., *Tetrahedron*, 43, 1197, 1987.
79. Warman, J. M., De Hass, M. P., Grätzel, M., and Infelta, P. P., *Nature (London)*, 310, 306, 1984.
80. Kunst, M., Beck, G., and Tributsch, H., *J. Electrochem. Soc.*, 131, 954, 1984.
81. Sahyun, M. R. V., *Chem. Phys. Lett.*, 112, 571, 1984.
82. Healy, T. W. and White, L. R., *Adv. Colloid Interface Sci.*, 9, 303, 1978.
83. Alfassi, Z., Bahnemann, D., and Henglein, A., *J. Phys. Chem.*, 86, 4656, 1982.
84. Howe, R. and Grätzel, M., *J. Phys. Chem.*, 89, 4495, 1985.
85. Kölle, U., Moser, J., and Grätzel, M., *Inorg. Chem.*, 24, 2253, 1985.
86. Rothenberger, G., Moser, J., Grätzel, M., Serpone, N., and Sharma, D. K., *J. Am. Chem. Soc.*, 107, 8054, 1985.
87. Boehm, H. P., *Discuss. Faraday Soc.*, 52, 264, 1971.
88. Nakato, Y., Tsumura, A., and Tsubomura, H., *J. Phys. Chem.*, 87, 2402, 1983.
89. Wilson, R. H., *J. Electrochem. Soc.*, 127, 228, 1980.
90. Yesodharan, E., Yesodharan, S., and Grätzel, M., *Sol. Energy Mater.*, 10, 287, 1984.
91. Nosaka, Y. and Fox, M. A., *J. Phys. Chem.*, 90, 6521, 1986.
92. Weyl, W. A. and Förland, T., *Ind. Eng. Chem.*, 42, 257, 1950.
93. McTaggart, F. K. and Bear, J., *J. Appl. Chem.*, 5, 643, 1955.
94. Moser, J., Gallay, R., and Grätzel, M., *Helv. Chim. Acta*, 70, 1596, 1987.
95. Albery, W. J. and Bartlett, P. V., *J. Electroanal. Chem.*, 57, 139, 1982.
96. Houlding, V., Geiger, T., Kölle, U., and Grätzel, M., *J. Chem. Soc. Chem. Commun.*, 682, 1982.
97. Cuendet, P. and Grätzel, M., *Photochem Photobiol.*, 36, 203, 1982.
98. Geiger, T., Nottenberg, R., and Pélaprat, M. L., *Helv. Chim. Acta*, 65, 2507, 1982.
99. El Murr, N., *Transition Met. Chem.*, 6, 321, 1981.
100. Grätzel, M. and Frank, A. J., *J. Phys. Chem.*, 86, 2964, 1982.
101. Rossetti, R. and Brus, L., *J. Am. Chem. Soc.*, 104, 2321, 1982.
102. Rossetti, R., Beck, S. M., and Brus, L., *J. Am. Chem. Soc.*, 106, 980, 1984.
103. Chandrasekaran, K. and Thomas, J. K., *J. Chem. Soc. Faraday Trans. 1*, 80, 1163, 1984.
104. Brown, G. T. and Darwent, J. R., *J. Chem. Soc. Chem. Commun.*, 93, 1985.
105. Moser, J. and Grätzel, M., *J. Am. Chem. Soc.*, 105, 6547, 1983.
106. Brown, G. T. and Darwent, J. R., *J. Chem. Soc. Faraday Trans. 1*, 80, 1631, 1984.
107. Miller, J. R., Beitz, J. V., and Huddeston, R. K., *J. Am. Chem. Soc.*, 106, 5057, 1984.
108. Serpone, N., Sharma, D. K., Moser, J., and Grätzel, M., *Chem. Phys. Lett.*, in press.
109. Kuczynski, J. and Thomas, J. K., *Chem. Phys. Lett.*, 88, 445, 1982.
110. Grätzel, M., *Acc. Chem. Res.*, 14, 376, 1981.
111. Albery, W. J., Bartlett, P. N., Wilde, C. P., and Darwent, J. R., *J. Am. Chem. Soc.*, 107, 1854, 1985.
112. Henglein, A., *Ber. Bunsengens. Phys. Chem.*, 86, 241, 1982.
113. Kuhn, A. T. and Mortimer, C. J., *J. Electrochem. Soc.*, 120, 231, 1973.
114. Dimitrijevic, N. M., Li, S., and Grätzel, M., *J. Am. Chem. Soc.*, 106, 6565, 1984.
115. Erbs, W., Desilvestro, J., Borgarello, E., and Grätzel, M., *J. Phys. Chem.*, 88, 4001, 1984.
116. Chandrasekaran, K. and Thomas, J. K., *Chem. Phys. Lett.*, 99, 7, 1983.
117. Bourdon, J., *J. Phys. Chem.*, 69, 705, 1965.
118. Borg, W. E. and Hauffe, K. H., *Current Problems in Electrophotography*, Walter de Gruyter, Berlin, 1972.
119. Giraudeau, A., Fan, F. R. F., and Bard, A. J., *J. Am. Chem. Soc.*, 102, 5137, 1980; Gosh, P. K. and Spiro, T. G., *J. Am. Chem. Soc.*, 102, 5543, 1980; Dave-Edwards, M. P., Goodenough, J. B., Hamnet, A. J., Seddon, K. R., and Wright, R. D., *Faraday Discuss. Chem. Soc.*, 70, 285, 1980; Mackor, A. and Schoonman, J., *Recl. Trav. Chim. Pays-Bas*, 99, 71, 1980; Memming, R., *Surf. Sci.*, 101, 551, 1980.
120. Clark, W. D. K. and Sutin, D., *J. Am. Chem. Soc.*, 99, 4676, 1977; Tinnemans, A. H. A. and Mackor, A., *Recl. Trav. Chim. Pays-Bas*, 100, 295, 1981; Hamnet, A. J., Dave-Edwards, M. P., Wright, R. D., Seddon, K. R., and Goodenough, J. B., *J. Phys. Chem.*, 83, 3280, 1979.
121. Gleria, M. and Memming, R., *Z. Phys. Chem. N.F.*, 98, 303, 1975.
122. Matsumura, M., Mitsuda, K., Yoshizawa, N., and Tsubomura, H., *Bull. Chem. Soc. Jpn.*, 54, 692, 1981.
123. Watanabe, T., Fujishima, A., and Honda, K., *Energy Resources Through Photochemistry and Catalysis*, Grätzel, M., Ed., Academic Press, New York, 1983.
124. Kiwi, J., *Chem. Phys. Lett.*, 102, 379, 1983.
125. Kamat, P. V. and Fox, M. A., *Chem. Phys. Lett.*, 102, 379, 1983.
126. Rossetti, R. and Brus, L. E., *J. Am. Chem. Soc.*, 106, 4336, 1984.
127. Moser, J. and Grätzel, M., *J. Am. Chem. Soc.*, 106, 6557, 1984.

128. Kalyanasundaram, K., Vlachopoulos, N., Krishnan, V., Monnier, A., and Grätzel, M., *J. Phys. Chem.*, 91, 0000, 1987.

129. Desilvestro, J., Grätzel, M., Kavan, L., Moser, J., and Augustynski, J., *J. Am. Chem. Soc.*, 107, 2988, 1985.

130. Vrachnou, E., Vlachopoulos, N., and Grätzel, M., *J. Chem. Soc. Chem. Commun.*, 868, 1987.

131. Shimidzu, T., Iyoda, T., and Koide, Y., *J. Am. Chem. Soc.*, 107, 35, 1985.

132. Kamat, P. V., Chauvet, J. P., and Fessenden, R. W., *J. Phys. Chem.*, 90, 1398, 1986.

133. Riefkohl, J., Rodriguez, L., Romero, L., Sampoll, G., and Souto, F. A., *J. Phys. Chem.*, 90, 6068, 1986.

134. Hashimoto, T. and Sakata, T., *J. Phys. Chem.*, 90, 4474, 1986.

135. Kalyanasundaram, K., in *Energy Resources Through Photochemistry and Catalysis*, Grätzel, M., Ed., Acdemic Press, New York, 1983, 217.

136. Kalyanasundaram, K. and Grätzel, M., in *Chemistry and Physics of Solid Surfaces*, Vanselow, V. R. and Howe, R., Eds., Springer-Verlag, Berlin, 1984, 111.

137. Bard, A. J., *J. Photochem.*, 10, 59, 1979; *J. Phys. Chem.*, 86, 172, 1982.

138. Bard, A. J., *Science*, 207, 138, 1980.

139. Hyne, J. B., in *Sulfur: New Sources and Uses* (ACS Symposium Series 183), Raymont, M.E.D., Ed., American Chemical Society, Washington, D.C., 1982; *Chem. Tech.*, 628, 1982.

140. Borgarello, E., Kalyanasundaram, K., Grätzel, M., and Pelizzetti, E., *Helv. Chim. Acta*, 65, 243, 1982.

141. Bühler, N., Meier, K., and Reber, J.-F., *J. Phys. Chem.*, 88, 3261, 1984.

142. Thewissen, D. H. M. W., Timmer, K., Eenhorst-Reinten, M., Tinnemans, A. H. A., and Mackor, A., *Nouv. J. Chim.*, 7, 191, 1983.

143. Wagner, C. and Traud, W., *Z. Electrochem.*, 44, 391, 1938.

144. Spiro, M., *Chem. Soc. Rev.*, 15, 141, 1986.

145. Grätzel, M., Kalyanasundaram, K., and Kiwi, J., *Struct. Bonding*, 49, 37, 1982.

146. Kalyanasundaram, K., Micic, O., Grätzel, M., and Grätzel, C. K., *Helv. Chim. Acta*, 62, 2432, 1979.

147. Kawai, T. and Sakata, T., *Nature (London)*, 286, 474, 1980.

148. Sakata, T. and Kawai, T., *Nouv. J. Chim.*, 5, 279, 1981.

149. St. John, M. R., Furgala, A. J., and Sammells, A. F., *J. Phys. Chem.*, 87, 801, 1983.

150. Olivier, B. G., Cosgove, E. G., and Carey, J. H., *Environ. Sci. Tech.*, 13, 1075, 1979.

151. Pruden, A. L. and Ollis, D. F., *Environ. Sci. Technol.*, 17, 628, 1983; *J. Catal.*, 82, 418, 1983.

152. Barbeni, M., Pramauro, E., Pelizzetti, E., Borgarello, E., Grätzel, M., and Serpone, N., *Nouv. J. Chim.*, 8, 547, 1984.

153. Grätzel, C. K., Jirousek, M., and Grätzel, M., *J. Mol. Catal.*, 39, 347, 1987.

154. Hidaka, H., Kubota, H., Grätzel, M., Serpone, N., and Pelizzetti, E., *Nouv. J. Chim.*, 9, 67, 1985.

155. Ross, R. T., *J. Chem. Phys.*, 45, 1, 1966; Ross, R. T. and Calvin, M., *Biophys. J.*, 7, 595, 1967.

156. Bolton, J. R., *Science*, 202, 705, 1978.

157. Almgren, M., *Photochem. Photobiol.*, 27, 603, 1978.

158. Rehm, D. and Weller, A., *Ber. Bunsenges. Phys. Chem.*, 73, 834, 1969.

159. Balzani, V., Boletta, F., Gandolfi, M. T., and Maestri, M., *Top. Curr. Chem.*, 75, 1, 1978.

160. Borgaya, F., Challal, D., Fripiat, J. J., and Van Damme, H., *Nouv. J. Chim.*, 9, 721, 1985.

161. Getoff, N. et al., Proc. Int. Conf. Hydrogen Energy, Vienna, 1986.

162. Duonghong, D., Serpone, N., and Grätzel, M., *Helv. Chim. Acta*, 67, 1012, 1984.

163. Khan, M. M. T. et al., Proc. Int. Conf. Hydrogen Energy, Vienna, 1986.

164. Khan, M. M. T., Bhandwaj, R. C., and Jadhar, C. M., *J. Chem. Soc. Chem. Commun.*, 1690, 1985.

165. Naman, S. A., Aliwi, S. M., and Al-Emaru, K., *Nouv. J. Chim.*, 9, 687, 1985.

166. Fujishima, A. and Honda, K., *Nature (London)*, 237, 37, 1972.

167. Tributsch, H. and Gerischer, H., *Ber. Bunsenges. Phys. Chem.*, 72, 437, 1968.

168. Memming, R., *Prog. Surf. Sci.*, 17, 7, 1984.

169. Matsumura, M., Matsudaira, S., Tsubomura, H., Takata, M., and Yanagida, H., *Ind. Eng. Chem. Prod. Res. Dev.*, 19, 415, 1980.

170. Alonso, N., Beley, V. M., Chartier, P., and Ern, V., *Rev. Phys. Appl.*, 16, 5, 1981.

171. Vlachopoulos, N., Liska, P., McEvoy, A. J., and Grätzel, M., European Conf. Surface Science, Lucerne, Switzerland, 1987.

172. Heller, A., *Acc. Chem. Res.*, 14, 154, 1981.

173. Balzani, V., *Nato Adv. Study*, 214, 1, 1987.

174. Hasegawa, T. and de Mayo, P., *Langmuir*, 2, 302, 1986; Yonagida, S. and Mizumoto, K., *J. Am. Chem. Soc.*, 108, 647, 1986.

175. Okahata, Y., Lim, H. J., and Hachiya, S., *Makromol. Chem. Rapid Commun.*, 4, 303, 1986.

176. Tricot, Y. M. and Fendler, J. H., in *Homogeneous and Heterogeneous Photocatalysis*, (NATO ASI Series C), Vol. 174, Pelizzetti, E. and Serpone, N., Eds., D. Reidel, Dordrecht, Netherlands, 1986, 241.

177. **Kakuta, N., White, J. M., Campion, A., Fox, M. A., and Webber, S. E.,** *J. Phys. Chem.,* 89, 48, 1985.
178. **Eibner, A.,** *Chem. Z.,* 35, 735, 1911.
179. **Tamman, G.,** *Z. Anorg. Chem.,* 114, 15, 1920.
180. **Baur, E. and Perret, A.,** *Helv. Chim. Acta.,* 7, 910, 1924.
181. **Krasnovsky, A. A. and Brin, G. P.,** *Dokl. Akad. Nauk SSSR,* 147, 656, 1962.
182. **Gerischer, H.,** *Pure Appl. Chem.,* 52, 2649, 1980.
183. **Kalyanasundaram, K.,** *Sol. Cells,* 15, 93, 1985.

INDEX

Printed and bound by CPI Group (UK) Ltd, Croydon, CR0 4YY

22/10/2024

01777600-0015